# Palgrave Studies in Media and Environmental Communication

Series Editors
Anders Hansen
School of Media, Communication and Sociology
University of Leicester
Leicester, UK

Steve Depoe
McMicken College of Arts and Sciences
University of Cincinnati
Cincinnati, OH, USA

Drawing on both leading and emerging scholars of environmental communication, the Palgrave Studies in Media and Environmental Communication Series features books on the key roles of media and communication processes in relation to a broad range of global as well as national/local environmental issues, crises and disasters. Characteristic of the cross-disciplinary nature of environmental communication, the books showcase a broad variety of theories, methods and perspectives for the study of media and communication processes regarding the environment. Common to these is the endeavour to describe, analyse, understand and explain the centrality of media and communication processes to public and political action on the environment.



More information about this series at
http://www.palgrave.com/gp/series/14612

David Robbins • Diarmuid Torney
Pat Brereton
Editors

# Ireland and the Climate Crisis

palgrave
macmillan

*Editors*
David Robbins
School of Communications
Dublin City University
Dublin, Ireland

Diarmuid Torney
School of Law and Government
Dublin City University
Dublin, Ireland

Pat Brereton
School of Communications
Dublin City University
Dublin, Ireland

Palgrave Studies in Media and Environmental Communication
ISBN 978-3-030-47589-5 ISBN 978-3-030-47587-1 (eBook)
https://doi.org/10.1007/978-3-030-47587-1



Cover illustration: Dave G Kelly / gettyimages
Cover design: eStudioCalamar

This Palgrave Macmillan imprint is published by the registered company Springer Nature Switzerland AG.
The registered company address is: Gewerbestrasse 11, 6330 Cham, Switzerland

# Foreword

The role of countries like Ireland in addressing climate change is too often overlooked. Climate change, as everyone knows, is the great challenge of our generation. Yet it is too often framed as a problem for the world's largest countries to sort out. After all, they have contributed most to the problem and their leaders often have much to say on the topic at G7 meetings. The world's attention is rightly focused at the present moment on responding to the COVID-19 health and economic crisis, but responding to climate change and setting the world on course for a rebalanced relationship to nature is our most pressing challenge.

A collective response to a global problem requires active leadership by all countries. Ireland is important as a proud member of the European Union, which continues to play an indispensable leadership role, and given its remarkable soft power on the international stage, building on nearly a century of diplomacy and drawing on deep wells of goodwill from its decades as a non-aligned party and as a leader in development cooperation. Beyond these formidable assets, Ireland has an inspiring story to tell about its own recent experience of climate action.

My friend and colleague Mary Robinson said recently, "Feeling a complete inability to do anything—'This is too big for me, I give up'—that's no use to anybody". She was talking about how individuals and societies should react to the grim daily news about climate change, but I think this also applies to countries. No one country can avoid its responsibility by saying that its individual contribution to the problem is too small. Ireland, to its credit, is showing that it realises the role it can play goes far beyond its greenhouse gas emissions.

In the last few years, Ireland has moved from its self-declared status as a "laggard" to a point where it is on the cusp of a new era of climate ambition. The turning point may have been in its innovative Citizens' Assembly, which in 2017 considered the question "How the State can make Ireland a leader in tackling climate change" and in the process revealed a stunning consensus in favour of measures that politicians had long thought would be unpalatable to the electorate. One of the most notable recommendations of the Citizens' Assembly was a far-reaching pledge to place climate change at the centre of policymaking in Ireland. This high-level recommendation was translated by a special parliamentary committee on climate action into a radical proposal to overhaul Ireland's 2015 climate law.

The Citizens' Assembly, as a model of deliberative democracy, may be one of Ireland's most consequential gifts to the world on climate change (though not just on climate) so far. I have the honour of being co-president of France's *Convention Citoyenne pour le Climat*, a citizens' assembly on climate change inspired by the Irish experience, and I have been deeply moved to see the how its 150 participants have risen to the task, with seriousness and great dedication, of determining how their country should address this daunting challenge.

Yet Ireland's biggest contribution may be still to come. If Ireland transitions to a net-zero economy, it will mean tackling its emissions from agriculture and transport. Ireland is not unique in struggling to transform these sectors, but they are more dominant in Ireland's greenhouse gas emissions profile than in other countries. It will therefore have to put special effort into figuring out how, as a small, open, export-oriented country, it can develop a climate-friendly agriculture and food sector and how to reverse the legacy of decades of car-based urban planning.

The context in which Ireland—and the world—face these challenges has been upended in recent months by the COVID-19 pandemic. Some will seek to use this as an opportunity to slow progress towards building a more sustainable future, and government bandwidth for engaging on climate policies may be constrained. But climate change is a foreseeable global crisis which, like COVID-19, will have far-reaching and indiscriminate society-wide impacts. These impacts will not be reversible. Governments must learn from the current situation that they have a responsibility to manage this risk.

I am confident that Ireland can rise to these challenges. And when it does, it will provide a good-news story to inspire other countries facing similar challenges of their own. This, beyond the actual tonnes of greenhouse gases abated, could be Ireland's most decisive act towards the implementation of the Paris Agreement.

CEO, European Climate Foundation
Paris, France
May 2020 Laurence Tubiana

# Acknowledgements

The editors wish to acknowledge support from the book publication and the workshop support schemes run by the Faculty of Humanities & Social Sciences at Dublin City University. We also thank Professor Charlotte Burns of Sheffield University, and Professor Neil Carter of York University, for their advice in shaping the book. We thank the students of the MSc in Climate Change: Policy, Media and Society at DCU for their engagement and feedback in developing the book proposal. We are grateful to Conor McNally, a graduate of the MSc in Climate Change programme, for his help in preparing the manuscript. Finally, we thank all the chapter authors for their contributions, and their patience.

# Contents

# Notes on Contributors

**Edwin Alblas** is a PhD candidate in environmental law and regulation at the Sutherland School of Law, University College Dublin (UCD). His research is situated within the interdisciplinary Effective Nature Laws project, led by professor Suzanne Kingston (ERC funded). He is a graduate of Tilburg University (BA, LLM) and KU Leuven (LLM). Prior to starting his PhD, Edwin worked in the EU department of the Dutch Ministry of Interior and Kingdom Relations. In UCD, Edwin lectures on environmental law and policy, environmental politics and regulatory governance, at both undergraduate and postgraduate levels. His works have been published in several international peer-reviewed journals, in the fields of both (environmental) law and political science, including *European Union Politics*, *Review of European, Comparative and International Environmental Law* and *Theory and Practice of Legislation*.

**John Barry** is Professor of Green Political Economy in the School of History, Anthropology, Philosophy and Politics and Director of the Centre for Sustainability, Equality and Climate action at Queen's University Belfast. His areas of research include green moral and political theory; green, post-growth and heterodox political economy; the politics, policy and political economy of climate breakdown and low-carbon energy transitions; engaged research; the greening of citizenship and civic republicanism. He is also a recovering politician and indifferent cook.

**Maria Barry** is an assistant professor in DCU's School of STEM Education, Innovation and Global Studies where she teaches and researches in the areas of history and citizenship education. She is also a member of DCU's Centre for Human Rights and Citizenship Education. Previously, Maria has worked as a post primary teacher and development education manager in Trócaire (NGDO). Scholarly interests include global citizenship education, classroom practice, student voices, climate change education and education for social justice.

**Pat Brereton** has worked at the University of Bedfordshire in UK where he secured his PhD, before returning to Ireland and DCU in 2001. His research is focused on all aspects of eco-cinema and environmental communication as well as new media studies. His books include *Hollywood Utopia: Ecology in Contemporary American Cinema* (2005), *Smart Cinema, DVD Add-ons and New Media Audiences* (2012), *Environmental Ethics and Film* (2016) and *Environmental Literacy: New Digital Audiences* (2019). He teaches on the new Masters in Climate change at DCU and remains active in researching across all areas of environmental communication, most recently being co PI on an EPA-funded research projects looking at the Irish Citizens' Assembly and Climate Change.

**Patrick Bresnihan** is a lecturer in the Department of Geography at Maynooth University. He works across the interdisciplinary fields of political ecology, science and technology studies and environmental humanities. He has written extensively, including articles on water infrastructure, urban commons in post-crash Dublin and the poetics of John Clare. His book, *Transforming the Fisheries: Neoliberalism, Nature and the Commons* (2016), won the Geography Society of Ireland Book of the Year in 2018.

**Sabrina Dekker** holds a PhD from University College Dublin's School of Architecture, Planning and Environmental Policy, a Master of Public Affairs from Sciences Po Paris, a Masters in Public Policy from the Lee Kuan Yew School of Public Policy at the National University of Singapore and a BA from the University of British Columbia. Dekker joined DCU in July 2018 as a postdoctoral researcher for the Irish Climate Policy Evaluation project. Prior to this she was a climate change researcher with Codema, developing the Climate Change Action Plans for the four Dublin local authorities. She has experience in renewable energy, community and economic development, environmental policy, and public health.

**Declan Fahy** is an associate professor at DCU's School of Communications where he researches the public communication of science, environment and sustainability. His work also examines journalists and scientists who are influential public intellectuals, a topic he explored in his 2015 book, *The New Celebrity Scientists: Out of the Lab and Into the Limelight*. He is on the editorial advisory board of *Public Understanding of Science*, the editorial board of *Journal of Science and Popular Culture* and the editorial board of *Environmental Communication*.

**Lorna Gold** is a social scientist and climate activist living and working in Ireland. She teaches in the Department of Applied Social Studies in Maynooth University. She previously worked as Head of Policy and Advocacy with Trócaire, the Irish Catholic Development Agency, for almost 20 years. She holds a PhD in Economic Geography from the University of Glasgow. She is vice-chair of the board of the Global Catholic Climate Movement and is a member of the Irish government advisory group on the National Climate Dialogue. Her most recent book *Climate Generation – Awakening to our Children's Future* tells her personal story of waking up to the ecological emergency as a mother, academic and activist. Her research interests include Ireland's just transition, civil society movements and the role of faith groups in climate activism.

**Peadar Kirby** is Professor Emeritus of International Politics and Public Policy, University of Limerick, and holds a PhD from the London School of Economics. Since his retirement in 2012, he has held positions as adjunct professor in the Faculty of Political Science, University of Iceland, Reykjavik, and adjunct professor to the Network on Politics, Power and Society at Maynooth University. He has written widely on Irish and Latin American politics and political economy, on globalisation and on vulnerability/resilience. His latest book, *Karl Polanyi and Today's Global Crisis: Transforming Market Society in the Era of Climate Change* to be published by Bloomsbury in 2020. In 2018, *The Political Economy of the Low-Carbon Transition: Pathways Beyond Techno-optimism*, co-authored with Tadhg O'Mahony, was published by Palgrave Macmillan. He is co-editor with Ernest Garcia and Mercedes Martinez-Iglesias of *Transitioning to a Post-Carbon Society:*

*Degrowth, Austerity and Wellbeing* (Palgrave Macmillan, 2017). He is a resident of Cloughjordan ecovillage, Co. Tipperary.

**Conor Little** is a lecturer in the Department of Politics and Public Administration at the University of Limerick. His research interests include representative politics, public policy and the comparative politics of climate change. His research has been published in journals including *Environmental Politics*, *Irish Political Studies*, *Party Politics*, *Scandinavian Political Studies* and *West European Politics*. He has co-edited a symposium of *Irish Political Studies* on *The Politics of Climate Change* (2017) and a special issue of *Environmental Politics* on *Climate Politics in Small European States* (2019). He is developing the Irish Policy Agendas Project as part of the international Comparative Agendas Project network. He holds a PhD from the European University Institute in Florence and has worked at University College Dublin, Keele University and the University of Copenhagen.

**Benjamin Mallon** is Assistant Professor in Development and Intercultural Education in the School of STEM Education, Innovation & Global Studies in the Institute of Education, Dublin City University. He researches and teaches in the area of Global Citizenship Education, with a particular focus on pedagogical approaches which address conflict, challenge violence and support the development of peaceful societies. His current research projects explore the role of in-service teacher education in supporting teachers to develop their practices in Climate Change Education. Ben is a member of DCU's Centre for Human Rights and Citizenship Education and he sits on the Editorial Board of Policy and Practice: A Development Education Review.

**Damian McIlroy** is a PhD researcher in Queen's University Belfast, in the Department of History, Anthropology, Philosophy and Politics. His research interests include Green Political Theory, Marxist Critical Theory, Eco-Marxism, Eco-socialism, Just Transition, Climate Breakdown, Green Republican Theory, Post-Growth and Co-operative Economics, Devolution and Democracy, Trade Unionism and Labour Movements. His PhD research is aimed at furthering our understanding of the conditions necessary to successfully implement a Just Transition in Ireland. The project has a specific focus on the structural crisis of capitalism, its relationship with climate breakdown and the transformative potential of trade union agency for a place-based (city/region) Just Transition, in Northern Ireland and the Republic of Ireland.

**Sinéad Mercier** is a consultant on just transition, climate change law and policy. She has previously worked as climate advisor to the Seanad Civil Engagement Group on the Joint Oireachtas Committee on Climate Action, with the National Economic and Social Council, the Irish Green Party, Philip Lee law firm, Irish Congress of Trade Unions and Amnesty International. She has been involved in climate activism with various groups including the Dublin Ecofeminists, SIPTU, Young Worker's Network, Not Here Not Anywhere and Friends of the Earth. She is applying a human rights framework to climate action, climate justice and just transition principles with the Irish Human Rights and Equality Commission. A recent publication is 'Men Who Eat Ringforts' with Eddie Lenihan, Michael Holly and Askeaton Contemporary Arts.

**Trish Morgan** is an assistant professor in the School of Communications at Dublin City University, where she lectures on the Multimedia and Emerging media undergraduate and postgraduate degrees. She is an interdisciplinary scholar who combines theories from media and communications with perspectives from geography. Her central research interest is in the communication of environmental issues through theory, geographical and practice-based approaches. Trish's distinctive research has been published in high-ranking journals in media/communications and geography. She has been the principal investigator on two research projects funded by the Environmental Protection Agency of Ireland: *Going Green Digitally? Environmental crisis, consumption patterns and the evolving role of media* (2014–2016), and Sensing our world: *How digital media cultural practices can contribute to changing social norms around consumption (2019–2020).*

**Sadhbh O Neill** is a PhD candidate at the School of Politics and International Relations at University College Dublin researching ethics and carbon trading. She is a policy advisor to Stop Climate Chaos, a coalition of civil society organisations that advocate for climate action, and former policy advisor to the Joint Oireachtas Committee on Climate Action as well as spokesperson for Climate Case Ireland. She holds a BA from Trinity College Dublin in Philosophy and English and an MA from University College Dublin in Philosophy and Public Affairs.

**Roderic O'Gorman** has been a lecturer in the School of Law and Government in DCU since 2012, where he teaches Climate Change Law, Planning & Development Law and EU Law. His research interests are in Environmental Rights, Comparative Constitutional Law, European Economic Law and Socio-Economic Rights. He served as a Green Party councillor on Fingal County Council between 2014 and 2020. In February 2020 was elected as a TD in the Dublin West constituency, and in June 2020 he was appointed Minister for Children, Disability, Equality and Integration.

**David Robbins** is an assistant professor in the School of Communications at Dublin City University. He continues to work as a journalist, having spent 25 years in national media before becoming an academic. He is the author of *Climate Change, the Press and Politics in Ireland* (Routledge 2019), and his work on media coverage of climate change has appeared in *Environmental Communication*. He is the co-founder of DCU's MSc in Climate Change: Policy, Media & Society.

**Gabriela Martinez Sainz** is an Ad Astra Fellow and Assistant Professor at the School of Education in University College Dublin researching and teaching on Children's Rights, Global Citizenship and Education for Sustainable Development. She holds a PhD in Education and a Master's Degree in Educational Research from the University of Cambridge where she conducted qualitative research on the intersection of human rights education, reflective practice and professional knowledge. As a researcher, she has been affiliated to the Centre for Human Rights and Citizenship Education at Dublin City University, the Brazilian Centre for Analysis and Planning CEBRAP, the Center for Governance and Human Rights at the University of Cambridge and also the Center for Socio-legal Studies of the University of Oxford. She is the author of textbooks and educational materials on issues of citizenship, democracy and human rights for secondary schools.

**John Sweeney** is an Emeritus Professor from Maynooth University and is a graduate of the University of Glasgow. He has taught climatology at Maynooth and a number of universities in North America and Africa and has published over 100 scientific papers and authored/edited/co-authored 4 texts on various aspects of Irish climate. He has served as President of the Irish Meteorological Society, the Geographical Society of Ireland and An Taisce, the National Trust for Ireland as well as being the Irish representative on a number of European academic bodies. He has been involved in various international research projects and has led several national research projects examining aspects of climate change in Ireland. Professor Sweeney contributed to the reports of the Intergovernmental Panel on Climate Change, which was awarded the Nobel Peace Prize in 2007. He is also a regular contributor to print and broadcast media on matters related to climate change science and policy.

**Diarmuid Torney** is an associate professor in the School of Law and Government at Dublin City University. His research focuses on comparative and global politics of climate change, environment and energy. He is author of *European Climate Leadership in Question: Policies toward China and India* (MIT Press, 2015) and co-editor of *European Union External Environmental Policy: Rules, Regulation and Governance Beyond Borders* (Palgrave, 2018). He is the chair of Future Earth Ireland, the national committee convened by the Royal Irish Academy of Future Earth, a ten-year global research initiative to develop knowledge for responding effectively to the risks and opportunities of global environment change. He served as a member of the Expert Advisory Group to the Citizens' Assembly for its consideration of the topic "How the state can make Ireland a leader in tackling climate change".

**Fionnuala Waldron** is Cregan Professor of Education and Professor Emerita of Dublin City University. Prior to becoming a teacher educator in the late 1990s, Fionnuala worked as a primary teacher. Her areas of interest in teaching and research include history education, human rights and citizenship education and climate change education. In 2004, she established the Centre for Human Rights and Citizenship Education in collaboration with Amnesty Ireland. The Centre focuses on research, teaching and learning in the fields of human rights education, global citizenship education and climate change education. From 2009 to 2016, Fionnuala was Dean of Education in St Patrick's College, Drumcondra, and from 2015 to 2016, she held the position of Interim Executive Dean of the DCU Institute of Education. Fionnuala has publications in the following areas: teacher education, climate change education, human rights education, citizenship education, history education and nineteenth-century Irish history.

**Clare Watson** has a social work background and over the years has worked on many social, environmental and community-based projects.

She is co-author of the book *Campaigns and How to Win Them* (1997), author of the blog *Chasing Hubcaps* (2013) looking at the influence of human psychology and behaviour on people's reaction to climate change and co-author of an RTE Audience Council report analysing RTE's coverage of climate change (2014). Clare has experimented with low-carbon living for the past 20 years. In 2018, she completed PhD research in MaREI, UCC, on an EPA-funded project looking at why people are finding it so difficult to respond to climate change, and in particular the benefits of, and barriers to, community energy. She is working as MaREI Engagement Support Officer and, as part of a MaREI engaged research team, is supporting the development of, and leading a case study on, Dingle Peninsula 2030.

# List of Figures

# List of Tables

SECTION I

# Introduction and Context

# Introduction

*David Robbins, Diarmuid Torney, and Pat Brereton*

## Context and Rationale

The message from the natural sciences on climate change has been consistent for over 30 years now: humans are causing the planet to warm, and we have a limited time to slow, stop, and possibly reverse this process. Although the impacts have become more dramatic and the projections more dire, the core information has remained constant: this is bad; act now or it may be too late. A key question follows from this, and it is one that most of us have been scratching our heads about for almost 30 years too. Why do societies not take the action required? Why do they not adopt the measures so clearly spelled out by the science?

D. Robbins (✉) • P. Brereton
School of Communications, Dublin City University, Dublin, Ireland
e-mail: david.robbins@dcu.ie; pat.brereton@dcu.ie

D. Torney
School of Law and Government, Dublin City University, Dublin, Ireland
e-mail: diarmuid.torney@dcu.ie

 
D. Robbins et al. (eds.), *Ireland and the Climate Crisis*, Palgrave Studies in Media and Environmental Communication,
https://doi.org/10.1007/978-3-030-47587-1_1

Of course, there are many answers to this question, but at the heart of all of them is the notion that climate change has transformed from being solely a matter of physics to becoming a social, political, and general societal problem. The reason for inaction does not simply lie in a lack of scientific information, but rather because, as a so-called wicked problem, climate change discourses need to engage with and become moderated through a broad range of communication strategies, alongside feeding off various political, economic, policy, psychological, educational, and other inter-connecting social movements. It is helpful, as this volume illustrates, to re-focus on the "soft sciences" in order to expose and examine the growing social and political forces that help or hinder action on climate change.

The historical importance of scientific studies in situating the danger of climate change is affirmed by the opening chapter on the impacts of climate change on Ireland by John Sweeney, a contributing author of the Intergovernmental Panel on Climate Change (IPCC) Fourth Assessment Report, who provides the key scientific information which the rest of the book takes as its starting point. Furthermore, the role of climate modelling and engineering, for example, will remain an essential component in developing responses to climate change, and the work of atmospheric and oceanic science must continue. We are suggesting, however, that the reasons for inaction on climate change cannot simply be found in the physical sciences. It is to the social sciences and humanities we must also turn in order to help find our way out of the predicament identified by the physical sciences so many years ago.

We began to think about ways in which a focus on social arenas such as policy, media, and politics could be given more prominence in debates around climate change, and it was from these discussions that the idea for this book emerged. It was conceived as a way to bring together a range of subject experts to examine how Ireland's social responses to climate change have been shaped and how these responses compared internationally.

We hope the volume will be of particular value to Irish readers, but can others learn from the Irish case? We believe so. Ireland is interesting, both because it shares features with other territories and because it has some unique aspects as well. Ireland is a small, politically stable state in which the social conditions should be conducive to collective action. Yet such action has so far failed to appear at anything like the required scale or speed, and the reasons why are instructive to similar jurisdictions. Ireland's status as a peripheral country is also a point of commonality with many others around the world which orbit a larger country or bloc.

In Ireland, politicians often debate whether the country should orient itself towards the US or the European Union (EU), to "Boston or Berlin" as it is sometimes expressed. Ireland is also interesting because of this tension between these two centres of gravity, and both European and American readers can learn from the Irish experience. Ireland's media system, popular culture, politics, and worldview owe something to both federations. In the wake of Brexit, and of the rise of populism in the US and elsewhere, the way in which Ireland orients itself politically and culturally is also of wider interest.

## Overview of the Book

The nature of the Irish experience in responding to the challenges posed by climate change is described and critically analysed in detail in this book. Chapter 2, "Climate Change in Ireland: Science, Impacts and Adaptation", by climate scientist John Sweeney, sets out the scientific context for the rest of the book. Ireland's rainfall pattern, exposure to drought and to both coastal and river flooding, has been increased by climate change. A two-metre storm surge could expose Ireland to a €1 billion insurance liability as a result of flooding. Biodiversity has been severely affected, and biodiversity loss will increase. Meanwhile, diseases previously seen only in more southerly latitudes will affect crop yields in Ireland.

In Chap. 3, "Ireland's Policy Response to Climate Change: An Historical Overview", Diarmuid Torney provides a critical historical overview of the policy responses to climate change, from the Celtic Tiger era through the economic downturn and the subsequent return to growth. Ireland is constrained by a number of circumstances in its ability to formulate a cohesive climate response, including fragmented governance systems and low population density combined with dispersed housing patterns. However, "Ireland enters the third decade of the 21st century with surprising momentum", the author states, noting developments such as the Citizens' Assembly on climate change, the Joint Oireachtas Committee on Climate Action, and the government's 2019 Climate Action Plan. Whether this represents a real transformation in Ireland's approach to dealing with climate change or whether it is a temporary deviation from the historical pattern of policy and political inertia remains to be seen, he concludes.

The scientific and policy context having been established and examined in the first section of the book, the second section considers in more detail specific aspects of climate law, politics, litigation, and local government. In Chap. 4, "Climate Litigation, Politics and Policy Change: Lessons from Urgenda and Climate Case Ireland", Sadhbh O Neill and Edwin Alblas describe and analyse the use of litigation as a means of compelling governments to implement action on climate change. When the political system proves incapable of producing adequate measures, they argue that the courts provide a means for citizens effectively to bypass politics. The authors outline the landmark Urgenda case in the Netherlands, in which citizens won a case against their government for failing to protect them against the impacts of climate change. They analyse an Irish case inspired by the Urgenda victory: the suit by environmental non-governmental organisation (NGO) Friends of the Irish Environment, claiming that Ireland's 2017 National Mitigation Plan was insufficient to meet commitments made by Ireland under the Paris Agreement of 2015. Although the case was lost in the High Court, the authors argue that such legal activism can shape the political discourse and the policy agenda.

Roderic O'Gorman (Chap. 5, "Climate Law in Ireland: EU and National Dimensions"), who was elected as a Green Party TD in the February 2020 general election in the course of writing his chapter and appointed Minister for Children, Disability, Equality and Integration in June 2020, examines EU and national climate change legislation and measures Ireland's national climate law of 2015 against legislative best practice internationally. Ireland's response does not compare favourably. When the Climate Action and Low Carbon Development Act was finally passed in 2015, "the law's ambition and design fell well short of good practice", says O'Gorman. The outline of a Climate Action (Amendment) Bill was published shortly before the dissolution of the 32nd Dáil (parliament) in January 2020 and awaits enactment. This addresses many of the shortcomings of the 2015 Act and can be thought of as "climate action for slow learners".

Moving to climate politics, Conor Little (Chap. 6, "The Party Politics of Climate Change in Ireland") charts and analyses climate change as a party political issue, noting that it has traditionally been a low-priority consideration. Two instances of higher salience for climate change are recorded—2006–2007 and 2019. In both periods, Ireland's economic growth was strong and levels of unemployment were low, international events provided a media focus, and there was electoral competition involving the larger parties and the Green Party. Little considers policy areas such as agriculture, transport, carbon tax, and energy, examining party positions on each. He concludes that party political attitudes to climate change have been characterised by "low-salience consensus", but that this may change.

The response at local government level in Ireland has also been slow and piecemeal, argues Sabrina Dekker in Chap. 7, "Responding to Climate Change: The Role of Local Government in Ireland". The author considers barriers to climate action imposed by local government and planning legislation and examines the ease of access to public representatives, a particular feature of Ireland's political culture. However, some local authorities have shown leadership through the adoption of smart cities initiatives. A growing role for Climate Action Regional Offices can also help local authorities strengthen climate resilience in their areas.

The third section of the book considers Ireland's social, visual, cultural, and media responses to climate change. In Chap. 8, "Ecological Modernisation, Irish-Style: Explaining Ireland's Slow Transition to Low-Carbon Society", Declan Fahy suggests that Ireland's "confused and contradictory" responses to climate change make more sense when understood in the context of eco-modernism, a concept of sustainable development in which economic growth is considered compatible with environmental protection. However, to be eco-modern, one must first be modern, and Fahy argues that Ireland as a late moderniser was a poor fit for eco-modern policies initially. He examines the role of agriculture in shaping—and constraining—Ireland's approach to sustainable development. Ireland can be categorised as a "weak eco-moderniser", with a "calculated ambivalence over ecological issues". There are signs of movement towards a stronger form of eco-modernism recently; however, Ireland has not yet had the kind of broad social and political debate necessitated by the challenges of climate change.

Trish Morgan (Chap. 9, "Challenges and Potentials for Socio-Ecological Transformation: Considering Structural Aspects of Change") considers structural barriers to Ireland's transition to a sustainable future. She uses the theoretical lenses of political economy, political ecology and field theory to examine the inherent tensions between economic growth and sustainability. The notion of spatial and temporal "fixes"—that during an economic crisis, capital is not lost, but merely moved around in space and time—is also presented as a means of understanding Ireland's financial crisis in 2008. Morgan also analyses the media as a site of cultural production, arguing that they are compromised in communicating ideas, practices, and imaginaries by commercial imperatives. The author presents a case study of TV coverage on the national broadcaster Radio Telefis Éireann (RTÉ) during 2013–2014 concerning the release of the 5th Assessment Report of the IPCC, noting that advertisements "bundled" with news coverage undermined the message of the IPCC report.

The media are the source of much of the public's information on scientific topics such as climate change, and the levels and nature of climate coverage by journalists are the subject of Chap. 10, "Climate Change and the Media" by David Robbins. International climate coverage is growing but remains relatively low. It peaks at times of UN climate conferences and major scientific reports and then fades back to the margins. Ireland's coverage follows this pattern but exhibits some unique characteristics. The economic crisis of the late 2000s, which coincided with the term in government of the Green Party 2007–2011, the presence of a powerful agriculture lobby, and the precarious financial state of the indigenous media sector have all influenced coverage levels in Ireland. The nature of Irish coverage is shown to be conservative, with eco-modern discourses and political framings dominant. Ireland has even adopted its own brand of scepticism, involving sarcasm and dismissiveness rather than outright denial of climate science.

The response to climate change of Ireland's visual media—in the form of eco-cinema—is the focus of Pat Brereton's contribution (Chap. 11, "Cultural and Visual Responses to Climate Change: Ecological Reading of Irish Zombie Movies"). Brereton examines low-budget zombie movies such as *Dead Meat* (2004) as a site of environmental discourse, arguing that such films can engage new, younger audiences with the topic of environmental degradation and can use the creative imaginary of horror to encourage more engagement with environmental issues. Brereton argues that these "horror flicks" can help puncture the romanticisation of food, land, and the stewardship role of farmers and notes how the Irish hurley, which features prominently in *Dead Meat*, has been decoded as "the new chainsaw" in the eco-cinema literature.

The fourth section of the book presents various perspectives and analysis of more citizen-led climate initiatives. The role of low-carbon communities such as ecovillages, community energy initiatives, and activism from civil society is examined, alongside approaches to climate education and the Just Transition movement. In Chap. 12, "Community Engagement and Community Energy", Clare Watson analyses the community energy sector in Ireland and charts the progress of local energy and low-carbon transition projects, focusing on three in particular: the Templederry Community Windfarm in County Tipperary, the Aran Islands Energy Co-op in County Galway, and the Dingle 2020 initiative in County Kerry. Barriers to community projects that were identified in the late 1980s—lack of feed-in tariffs, difficulties in gaining planning permission, securing investment finance, and obtaining access to the grid—still persist and must be addressed. Ireland's All-of-Government Climate Action Plan of 2019 is an encouraging development, she concludes.

Education can play a key role in climate change mitigation and adaptation, say Fionnuala Waldron, Ben Mallon, Maria Barry, and Gaby Martinez Sainz (Chap. 13, "Climate Change Education in Ireland: Emerging Practice in a Context of Resistance"), but climate change education is a contested space. The Irish education system has focused on promoting environmental citizenship and climate action at individual and school level rather than on system critique or collective political action. Teaching climate change in the classroom can be challenging for teachers, and they must be supported to develop understandings of the complex issues surrounding climate action and climate justice, the authors argue. Climate change education which does not focus exclusively on "soft", apolitical approaches can "offer a valuable framework for educators to support the knowledge, understanding and collective action of children, young people and wider society", they conclude.

The role of trade unions in ensuring the transition to a low-carbon society is a just one is the focus of Chap. 14, "Climate Action via Just Transitions Across the Island of Ireland: Labour, Land and the Low-Carbon Transition", by Sinéad Mercier, Patrick Bresnihan, Damian McIlroy, and John Barry. The Just Transition perspective holds that social dialogue is necessary to ensure that the pay and conditions of workers involved in the transition are not adversely affected, and the authors argue that bargaining power through unionisation is vital. The Northern Ireland peace process has lessons around conflict management and transformation that are relevant to this transition process. The authors suggest that, when considering a just transition, Ireland's farmers should be thought of as the equivalent of oil and gas workers in other countries. Direct democratic state governance is needed to ensure that people, communities, and workers are not unfairly disadvantaged. Without such measures, support for any low-carbon transition becomes more elusive.

Lorna Gold provides a history of the civil society climate movements around the world and in Ireland (Chap. 15, "The Changing Faces of the Climate Movement in Ireland"), noting that NGOs and activists were focused on increasing the ambition of the United Nations Framework Convention on Climate Change (UNFCCC) process. After the failure of the UN climate summit in Copenhagen in 2009, climate activists disagreed about whether the UNFCCC process was the most effective focus of their efforts. Furthermore, while NGOs in northern countries emphasised global, science-driven efforts to reduce greenhouse gas (GHG) emissions, a new focus on climate justice emerged from the Global South. This spatially embedded justice perspective has in turn been superseded by the new youth climate movement, which foregrounds time and urgency. In Ireland, new groups have emerged and the climate movement is undergoing unprecedented change. Pre-existing power structures remain, even as new voices join the movement. The author concludes that "The important thing is not that everyone is saying the same thing, but that there is advocacy happening".

Lastly, Peadar Kirby, a resident in Ireland's only ecovillage, in Cloughjordan, County Tipperary, considers in Chap. 16 the role of ecovillages in modelling to the rest of society what low-carbon living looks like and reflects on the experience of the Cloughjordan ecovillagers in setting up their community. The role of participatory projects in tackling climate change has been recognised belatedly by the IPCC, and Kirby argues that "Ecovillages provide spaces where such transformative visions beyond business-as-usual scenarios can be promoted and implemented, and rich lessons learnt". Cloughjordan's ecovillage status rests on three pillars: ecological building standards, district heating, and food system. The ecovillage uses a bottom-up decision-making system inspired by ecology: the Viable Systems Model. Despite the emphasis on participation and consensus, divisions remain, the author notes. Cloughjordan was a pioneering project, and national regulations and practices have caught up with many aspects of its operation. The challenge for Ireland's first ecovillage now, argues the author, is to "remain ahead of the curve, learning and spreading the lessons of how vibrant, resilient and welcoming communities will be the seedbeds of a low-carbon Ireland".

## Cross-Cutting Themes

When considering the diversity of chapters in this book, and in reflecting on so many perspectives and accounts, certain motifs and commonalities emerge. The influence of agriculture on nearly every aspect of Ireland's response to climate change is immediately evident. It is mentioned in almost every chapter, and its shadow falls across social arenas from policy, media, and party politics to education, the Just Transition, considerations of eco-modernism, and even zombie movies. Ireland is an outlier when it comes to the contribution of agriculture to its national GHG emissions profile. Furthermore, land, farming, and farmers hold a near-sacred status in the national consciousness. Significantly reducing agricultural emissions is difficult without reducing livestock numbers and changing land use, and there has been a reluctance in many parts of Irish society to confront this issue and engage with farmers. If Ireland's transition to a low-carbon society and economy is to come about without deep social division, this is a conversation we need to have.

Related to the inertia, conflict avoidance and political long-fingering around agriculture is another cross-cutting feature: Ireland's weak institutional response to the challenges posed by climate change. Across governance, policy, party politics, local government, and even the media, the issue has been seen as low priority and has had low salience. Policy documents have been unambitious, legislation has been weak, and diplomatic capital has been spent in securing flexibilities for Ireland.

Environmental debates in Ireland have much in common with those in other jurisdictions, and the perennial tension between incremental versus radical change is another common feature across several social arenas. It emerges in the struggle between those who would like to see Ireland develop a stronger form of eco-modernism, those who wish to proceed more slowly and cautiously, and those who reject eco-modernism completely and call for more radical system change. This is evident in political and policy debates between "realos" and "fundis", and it is present among those who argue that the media should cover climate change as a systemic issue rather than continue with episodic coverage.

These debates are almost entirely concerned with mitigating the effects of climate change through various GHG emissions reduction policies and measures, while debates concerning adaptation to the effects of climate change are neglected. This one-eyed focus on mitigation is not unique to Ireland, but it is marked nonetheless. Across education, policy, and governance, attention is devoted very significantly to issues of mitigation. This is also true of media coverage, which is largely concerned with scientific reports and political posturing at the UN, EU, and national level. Considerable resources are being devoted to adaptation strategies at local and national policy level, but adaptation has not yet made its way into the wider social discourses around climate change.

Ireland's image—both its image of itself and the image perceived from the outside—is a recurrent theme. Ireland is seen as "green" because of its extensive pastureland and its relatively unindustrialised landscape. This green image has been used in government promotion campaigns related to food and tourism. As Trish Morgan points out, "Green is semiotically synonymous with environmental well-being and therefore, associates Ireland with positive connotations of nature". However, there is a marked disconnect between this green image and a reality in which Ireland's GHG emissions continue to rise and the country's water, soil, and biodiversity continue to be degraded by intensive agricultural practices. This disconnect has been identified and explored by filmmakers, and there is a role for eco-cinema in exploring a post-agricultural Ireland.

Ireland's responses to the particular challenges of climate change can read like a litany of failure. In chapter after chapter, failures of governance, of political courage, missed opportunities, chances for transformation forgone, are described and analysed. Yet in most cases, the chapters conclude by noting a change in what might be called the "mood music" around climate change. The final common theme of our book is hope. In governance, the 2019 Climate Action Plan is seen as potentially transformative; in media, the public broadcaster and private media outlets have responded to an increase in public interest; in civil society, new groups have re-energised and re-imagined climate activism. In other areas, around education, the Just Transition, and the visual media, awareness has grown and debates are ongoing. Just as Ireland was a late moderniser, and a late eco-moderniser, it is becoming a late adopter of climate action measures.

## The Influence of the Covid-19 Pandemic on Climate Debates

The arrival of the Covid-19 virus in Ireland in late 2019 or early 2020 prompted a swift and generally effective response by the Irish state. The resources of multiple agencies were mobilised, extensive lockdown measures were introduced, and some private assets were commandeered for public use. This wide-ranging intervention by the state shaped climate action debates in a number of ways. Firstly, it was evidence that concerted government action in the face of a crisis was effective and, perhaps more surprisingly, could enjoy widespread public support once the nature of the crisis was clearly communicated. Environmental campaigners began to ask why the government could not respond with similar urgency to the climate and biodiversity emergency.

Secondly, there was considerable public and media discourse concerning the improvements in environmental conditions during the lockdown. The reduction in transport activity and in manufacturing output meant that air and water quality improved, and carbon dioxide emissions are projected to have decreased significantly on a once-off basis. The lack of access to indoor exercise facilities meant that the public were forced to exercise outdoors, resulting in an increase in public awareness of the importance of green spaces in cities and the need for improved facilities for walkers and cyclists. The measures called for over many years by climate change activists—the reduction in private car use and greater support for modes of active travel, for instance—were inadvertently implemented and received widespread support.

Lastly, the debates about the economic cost of interventionist measures were changed by the pandemic. The Irish government instituted various wage-support and welfare schemes and pledged that post-Covid efforts to deal with the cost of these would not involve a return to the austerity measures that followed the 2008 economic crash. There was widespread social acceptance that human health was a higher-order concern than economic productivity or commercial activity, and the few public figures to call for an end to the Covid restrictions on economic grounds did not receive wide support. It became accepted that the government could borrow and spend as necessary in the face of a public health emergency. Again, environmentalists pointed to the parallels: if practically unlimited public funds could be made available to deal with a health emergency, why not for a potentially even more damaging climate and biodiversity one?

Other aspects of the Covid-climate comparison have also become evident. Having been somewhat maligned over previous years, experts and expertise were suddenly back in fashion. The science and crisis communications aspect of the pandemic may have lessons for communication concerning climate change. The emergence of hyper-local economies, encouraged by 2 km and then 5 km travel restrictions, could serve as exemplars of local resilience. The community spirit and solidarity evident in Ireland's response to the pandemic point to a resource that could be mobilised in support of climate action.

The pandemic has also posed significant risks for the climate and broader sustainability agendas. Interest in sustainability has historically tended to wane during economic downturns as a result of both limited bandwidth of the policy system and constrained resources. Against this backdrop, environmentalists in Ireland and around the world have sought to learn lessons from the recovery from the global financial crisis, widely seen as a missed opportunity to shape a green recovery. The need to "build back better" became the slogan of those seeking to integrate climate and sustainability into national Covid-19 recovery plans, for example, by calibrating stimulus plans to deliver sustainability objectives as well.

Against this uncertain and evolving backdrop, the core themes of the following chapters are more pertinent than ever. Ultimately, it remains to be seen whether we have left it too late or whether more recent public and political engagement with the issue will enable Ireland to dispense with its "climate laggard" reputation at last.

# Climate Change in Ireland: Science, Impacts and Adaptation

*John Sweeney*

## Climate Change and Ireland: Establishing the Science

Climate change has been described as a 'wicked' problem. This is a concept introduced almost half a century ago to describe categories of problems that are rooted in conflicting societal interests and featuring multiple stakeholders who do not share a consensus on what the problem is, far less how to address it (Rittel and Webber 1973). By this definition, climate change is what Levin et al. (2012) characterised as a 'super wicked problem'. It matches perfectly their four criteria: the time for addressing the problem is running out, competing agendas between the perpetrators and those seeking a solution exists, an effective central governance body is absent and an inability to see beyond short-term pain and long-term gain is evident. All of these compromise the ability of society and individuals within it to take effective action.

J. Sweeney (✉)
Maynooth University, Maynooth, Ireland
e-mail: John.Sweeney@mu.ie



D. Robbins et al. (eds.), *Ireland and the Climate Crisis*, Palgrave Studies in Media and Environmental Communication,
https://doi.org/10.1007/978-3-030-47587-1_2

Scientific knowledge alone cannot provide solutions to this 'wicked' problem. However, scientifically informed perspectives can provide the grounding for decisions to mitigate its impacts and adapt to the challenges it poses. Since its establishment in 1988, the Intergovernmental Panel on Climate Change (IPCC) has provided an objective, non-policy-prescriptive review of the state of the art in climate science research. This has proven to be an invaluable aid to policymakers seeking to reconcile international and domestic priorities. Successive reports from this body have sought to inform international initiatives such as the United Nations Framework Convention on Climate Change, the Kyoto Protocol and, most recently, the Paris Agreement. Interspersed with major Assessment Reports published every six years or so, Special Reports on key topics such as the Ocean and Cryosphere (IPCC 2019a, b, c), Land (IPCC 2019a, b, c) and Global Warming of 1.5°C (IPCC 2018) clarify key scientific issues for the benefit of international negotiators. A standardised methodology for the calculation of national greenhouse gas inventories is also produced by the Panel, updated most recently in 2019.

Such is the complexity of the earth system that an operational hardware model of its functioning will always remain impossible. Reliance on conceptual computerised models of the earth-atmosphere system has therefore been the principal vehicle for advancements in climate science. As computing power has increased, these have evolved from simple energy balance-type models to fully coupled earth system models which perform well when tested through hindcasting, that is, modelling past climate changes. This promotes confidence in their ability to project future climate scenarios and provides confirmation of the anthropogenic control of contemporary climate.

In the lead up to the next IPCC Assessment Report, considerable advances have been made in climate science. Apart from increased confidence in the models being developed, reductions in uncertainties in cloud feedback (Satoh et al. 2018) and in snow-albedo feedback are also being achieved (Thackeray and Fletcher 2016). A major focus of the next IPCC Assessment Report is likely to be on the issue of Equilibrium Climate Sensitivity (ECS), that is, the equilibrium change in global mean temperature that would result from a doubling of the atmospheric equivalent $CO_2$. No shift in the range of resultant temperature values as estimated by the IPCC has occurred between the First Assessment Report in 1990 and the Fifth Assessment Report in 2013. This is a crucial parameter for driving global climate models (GCMs), and the IPCC Fifth Assessment Report stated that 'there is high confidence that ECS is extremely unlikely to be less than 1°C and medium confidence that the ECS is likely between 1.5°C and 4.5°C and very unlikely greater than 6°C'. A narrowing of this range is likely to occur, with most estimates clustering around 3°C (Rahmstorf 2008). More recent work suggests that there is only a 4% probability of ECS falling outside the 1.5–4.5°C range (Cox et al. 2018).

One of the most beneficial aspects of the dramatic increase in computing power in recent years has been the ability to run global climate models (GCMs) multiple times, frequently from slightly different initial conditions. This has enabled a reduction in model uncertainties to be achieved by grouping outputs in what are known as ensemble runs. In addition, multiple runs can facilitate estimates of how likely given extremes are to occur with pre-industrial and present greenhouse gas concentrations. Thus, for example, climate change made Europe's heatwave of July 2019, which produced record-breaking temperatures of over 40 °C over large areas, up to 100 times more likely than would have been the case with pre-industrial concentrations of greenhouse gases in the atmosphere (World Weather Attribution 2019).

Due to the relatively coarse resolution of global climate models, their utility for policymakers is limited. Accordingly, a variety of downscaling techniques have been developed to provide the high-resolution scenarios that would enable decision makers to plan adaptation strategies at regional and local scales more effectively. In the case of Ireland, these initially involved establishing statistical relationships between current local climate variables and large-scale predictors, such as pressure fields, that can then be related to global climate model outputs in order to simulate future conditions (Sweeney et al. 2003, 2008; Fealy and Sweeney 2008). With further improvements in computing power, a dynamical approach has become dominant, whereby output from the GCM is used to drive a regional climate model (RCM) at higher spatial resolution. Using one model to drive another one inevitably introduces a cascade of uncertainty and so ensembles of both the GCM and the RCM are frequently incorporated into the outputs. What this means is that rather than running a single climate model, a suite of multiple model versions is run with each version of the model being slightly different from one another. This produces a range of results. The extent to which these cluster around a narrow or wide range gives a measure of uncertainty to the projections concerned.

The diversity of GCMs and RCMs employed in simulations has steadily increased over time and a range of projections for Ireland's future climate has emerged (McGrath et al. 2005; McGrath and Lynch 2008; Gleeson et al. 2013; Nolan 2015; Fealy et al. 2018). These have generally shown a reasonable degree of consistency in terms of temperature and rainfall, though uncertainties persist in terms of seasonal and spatial distributions. Rainfall remains the most difficult parameter to tie down.

Mean temperatures during the 1981–2010 standard averaging period are 0.5°C warmer at almost all Irish locations than during the comparable 1961–1990 period, a trend that will be further reinforced with the release of the 1990–2020 figures. Warming is presently occurring at a rate of 0.14°C/decade. By mid-century, a warming of 1–1.6°C is indicated (Fig. 1). This is slightly less than the global rise expected over the same period and reflects Ireland's oceanic location and thus its tendency to lag behind the trends established in more continental locations. But as a mid-latitude country, the Irish temperature change is likely to ultimately be close to the global mean figure. Hot days and warmer nights will accompany this warming, and a major reduction in frost frequency will occur, meaning that many coastal locations will often not experience frost at all during some winters. The growing season will extend by over a month at most locations, with consequences for ecology and agriculture.

Most models agree that Ireland will face significant decreases in rainfall during future spring and summers. But droughts are not prominent in the psyche of most Irish people. More accepted are the frequent occurrences when strings of frontal depressions cross the country and maritime air masses make for cool, disappointing summers. However, a newly developed 305-year record of rainfall for Ireland shows that Ireland is surprisingly drought prone, with seven major drought-rich periods since 1850 alone (Murphy et al. 2018). The work further demonstrates that recent decades are unrepresentative of the island's long-term drought climatology. Most models suggest that summer rainfall reductions will be most severe in the eastern half of the island, already a region where rising population, and thus rising consumer demand, has placed public water supplies in a precarious situation. By contrast, winter rainfall is expected to increase, especially in western parts, posing an increased threat of flooding (Fig. 1). The long-term record of seasonal rainfall is in line with model projections showing an upward trend in winter rainfall and a downward trend in summer rainfall, though considerable variability exists in short-term receipts.

A change in the frequency of extreme events is a characteristic harbinger of future climate change and the past decade in Ireland has provided an exemplary record. The last 10 years have been the wettest in 300 years

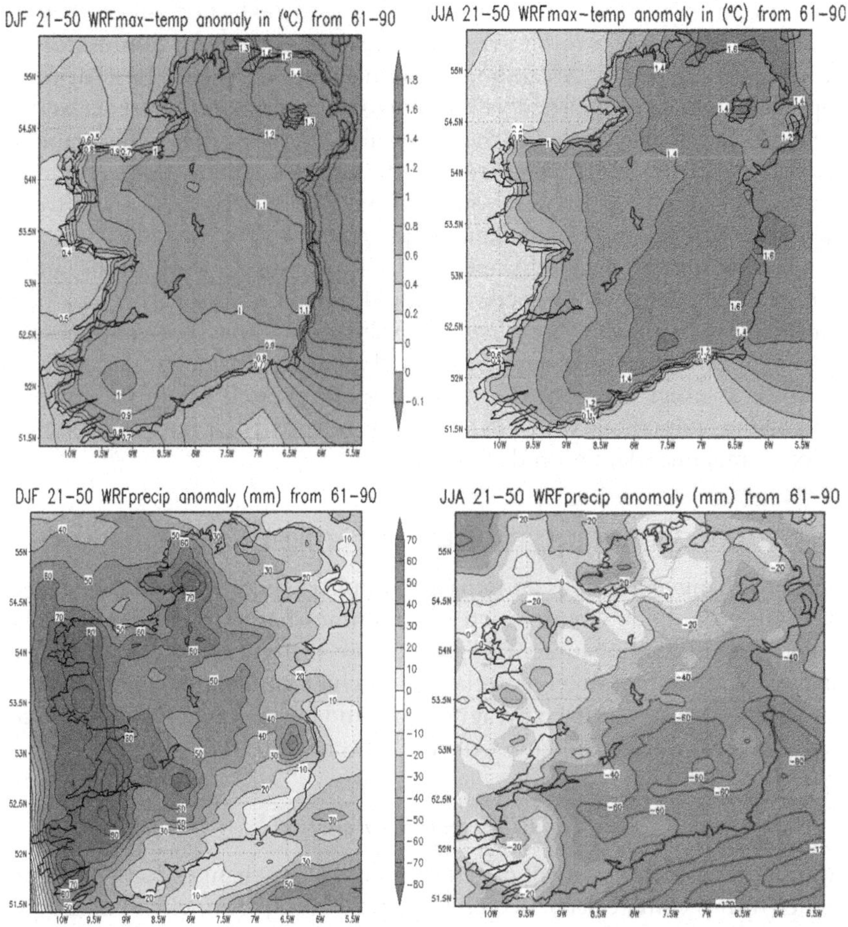


**Fig. 1** Projected temperature and rainfall changes in Ireland from 1961 to 1990 and 2021 to 2050. (Source: GCM: EC-Earth, RCP8.5, RCM: WRF, courtesy of ICARUS, Maynooth University)

with winter 2015/16 being the wettest winter on record for a majority of Ireland's weather stations. Winter 2013/14 was the stormiest season for at least 143 years for Ireland and Britain (Matthews et al. 2014), while summer 2018 was the driest on record at some south coast locations. These events are in line with model projections. Storms are projected to decrease in frequency as the jet stream moves north, but, when they do approach Ireland, to exhibit greater intensity as the energy content of the atmosphere increases. A candidate for this scenario might be Hurricane Ophelia which was regarded as the most severe storm to affect Ireland in 50 years. It was also classed as the easternmost Atlantic major hurricane on record. In 2019, Hurricane Lorenzo followed a similar track, reaching Category 5 status for a time, though weakening to become an extra tropical storm well before it made landfall in Ireland.

Uncertainty will always be a feature of future model projections, and it is important that such projections should not be used slavishly as a basis for adaptation policies. So while it is valid to suggest that century-scale extremes of drought or wetness may change to decadal scales under a business-as-usual scenario, the value of such research lies in its utility in identifying vulnerability and in exploring sensitivity of particular sectors prior to making adaptation decisions.

## Climate Change and Ireland: Projecting the Impacts

Climate change will act as an impact multiplier where vulnerabilities exist, through a complex nexus which is sometimes hard to disentangle. This makes for difficulties in adopting a narrow sectoral approach to impact analysis and adaptation. For Ireland, the chief vulnerabilities lie in three interrelated areas: biodiversity, agriculture and forestry, and water resources and the coastal environment (Coll and Sweeney 2013).

### *Biodiversity Impacts*

Many species in Ireland will experience significant changes to their ranges under future climate scenarios (Byrne et al. 2003; Donnelly et al. 2008). Species with highly specialised requirements will be susceptible to the largest range changes. The distribution of some, such as moss, liverwort and fern species, will contract. Cold-loving montane species will be particularly vulnerable since these species will not be able to move to higher altitudes and latitudes. Others will not be capable of migrating to new areas of suitable climate and habitat or adapting to new conditions. The maintenance and promotion of connectivity in the wider landscape is therefore vital to ensure that species can reach new areas of suitable climate space. The creation of sustainable landscape management approaches, such as by well-designed agri-environment measures, which enable natural processes to take place, and thereby increase the resilience of ecosystems, will be essential to facilitate this.

Some habitats will be particularly vulnerable. These include upland habitats, peatlands and coastal habitats (especially fixed dunes), where the additional threat of sea-level rise will be involved. Only 9% of the habitats protected under the European Union (EU) Habitats Directive have a 'favourable' conservation status in Ireland (NPWS 2013), and concern exists that climate change will accentuate their degradation. Coll et al. (2014, 2016) used climate envelope modelling to assess how habitats would respond to the projected climate changes and identified where and when losses of blanket bog and upland heaths would likely occur. A loss of 7.2% for blanket peats and 10–40% for wet heaths over the period 2031–2060 was projected (Fig. 2). Among the vulnerable habitats to respond most rapidly to climate change are coastal dune fields, and evidence exists that the recent changes in growing season have already resulted in vegetation of hitherto bare dune surfaces in many parts of Ireland (Jackson and Cooper 2011). A reduction of 80% in bare surfaces was identified as having occurred over the period 1985–2005, indicating

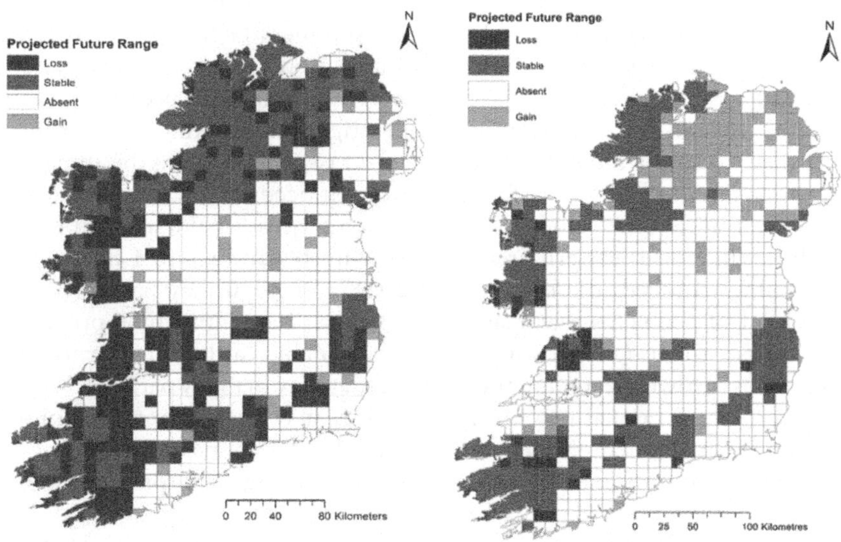


**Fig. 2** Projected loss of active blanket bogs (left) and wet heaths (right). (Figures show 10 × 10 km grid resolutions. Courtesy of Coll et al. 2014, 2016)

that a complex interplay between sea-level change, storminess changes and temperature is likely to characterise the evolution of these habitats in coming years. The declaration of a Climate and Biodiversity Emergency by the Dáil in May 2019 represented an acceptance of the interconnected nature of the two. As a threat multiplier, climate change interacts especially with the current policy of agricultural intensification to accelerate biodiversity losses. Independently of this, however, agriculture will itself face impacts from climate change, as well as adding to Ireland's contribution to the problem.

### *Agriculture and Forestry*

Using a downscaled GCM to drive crop simulation models for the middle and end of the century, Holden et al. (2003) suggested that cereal crops such as wheat and barley would continue to be viable, with their yields strongly influenced by the availability of precipitation in spring and summer. The major winner from climate change in Ireland will be maize which has the potential to become a valuable crop suitable for high-energy forage. Presently, maize silage is being substituted for grass silage and is proving more nutritious for livestock. This is likely to continue. By contrast, potatoes, highly sensitive to reductions in summer rainfall, can be expected to fare much worse. In the absence of irrigation, potato yields in eastern Ireland are projected to reduce considerably. To maintain present yields in eastern Ireland will require a significant irrigation contribution after mid-century. Ireland's most important crop, grass, will similarly be impacted by drier summers and more frequent drought occurrences. Here a marked east-west difference emerges with western areas maintaining present yields, while drier eastern areas of east Munster and south Leinster face difficulties. Indeed, the droughts of 2018 and 2020 may have presaged the kind of problems that may become more frequent as time proceeds. Of course, the longer growing season may enable cattle to be turned out onto grass growing earlier in the spring than today in many parts. However, greater winter rainfall may mean that stock may have to remain housed longer in spring than farmers would wish due to the risk of poaching on the wetter western soils. Problems with slurry storage and spreading in the wetter western parts of the country would also be expected to arise.

Increased $CO_2$ concentrations and warmer temperatures are likely to benefit Irish forest growth, with drier summers and changes in storm frequencies the main threats possibly negating some of the potential benefits

(Ray et al. 2007). But a major problem is likely to be increased incidences of pests and diseases. Insect pests in particular can respond rapidly to warmth. Many can rapidly multiply and go through more generations in warmer conditions. Among the most significant pest and disease threats are Green spruce aphid, Pine weevil (Williams et al. in prep), Great spruce bark beetle, European pine saw fly, Fomes, Phytopthera disease of alder and Honey fungus. A particularly visible problem in eastern Ireland at present is the Horse Chestnut Leaf Miner (*Cameraria ohridella*) which was first identified in the Republic of North Macedonia in 1984 and has since made its way north and west in Europe to attack horse chestnuts by mining into the leaf and rendering the tree seriously weakened in late spring (Fig. 3). Presently, the pest is moving west in Ireland from initial colonies established along the east coast in 2013 and, given the difficulties of eradicating a pest that overwinters in fallen leaves, will ultimately spread throughout the island. Of course, climate may be only a contributing factor to some of these problems, but given the present and future importance of forestry as a potential sink for $CO_2$, a long-term perspective is

**Fig. 3** Infestation by Horse Chestnut Leaf Miner (*Cameraria ohridella*) (Chapelizod, Dublin)

important, and pest and disease-resistant traits must be encouraged in tree-breeding programmes. Continued vigilance on imported wood products is also vital.

### *Water Resources*

For every 1 °C rise in temperature, the atmosphere can hold approximately 7% more water vapour, and so heavy rainfall events are likely to become more common in Ireland's future climate records. This is already occurring and exacerbating hazards associated with flooding. Much of the island's drainage (and water supply) infrastructure was constructed to cope with a past climate that is no longer appropriate to factoring in for calculation of flood risk. Of course, increased urbanisation, which by its nature consists of impermeable surfaces that deliver heavy rainfall quickly to streams and rivers, accentuates the problem, as do past planning mistakes which allowed development on flood plains which removed natural storage of flood waters resulting in significantly increased vulnerability downstream. Unsurprisingly, this combination of factors has rendered Ireland more exposed than most European countries, with flooding constituting the primary hazard associated with climate change for the foreseeable future. Some 300 risk areas have been identified. It is anticipated that up to €1 billion will be required over the coming decade to be spent on flood risk management in Ireland (Office of Public Works 2018a, b).

### *Coastal Environments*

With a coastline of 6500 km and a location astride the main Atlantic depression tracks, Ireland is highly vulnerable to changes in marine climate and to storm surges combining with higher sea levels. The crenulated shoreline of western Ireland is testament to the interplay between its geological skeleton of ancient rocks and the power of Atlantic wave attack. Carter (1990) quantified the high-energy waves involved at $10^{11}$–$10^{12}$ J/m$^2$/year. Expected changes in storminess will add further energy to such environments and will be compounded by a rise in sea level that will continue for centuries. Warming contributes to this in two ways. Firstly, the ocean column expands, and this has been responsible for about half of the global sea-level rise of the past 25 years. But secondly, as land ice melts and enters the ocean, this also raises sea level. The 5th IPCC Assessment

Report indicated that average losses in Greenland were contributing 215 Gt water to the global ocean during the period 1993–2009, some six times more than during the period 1992–2001. A contribution of 147 Gt per year was reported from the Antarctic after the turn of the twentieth century, again substantially enhanced over previous measurements. Unsurprisingly, therefore, sea-level rise has been accelerating and currently averages 3.4 mm per year. For Ireland, with a complex glacial history whereby some northern areas are still experiencing isostatic rebound from the release of their ice coverage, sea-level changes have varied geographically but threaten major urban areas such as Cork and Dublin. With 40% of the Irish population living within 5 km of the coast, and with many urban centres having expanded seawards from a more inland historic nucleus, vulnerability is obvious.

The 'soft' coastlines of eastern Ireland are particularly vulnerable. Rates of erosion along the dune coastline of Wexford are of the order of a metre per year, sometimes more (DELG 2001). Homes have already been lost. Using past insurance claims data for homes and businesses affected by river flooding, Flood and Sweeney (2012) estimated that a 2 m event (surge plus sea-level rise) could potentially generate insurance claims of over €1B nationally, with the bulk coming from Dublin and Cork (Table 1). Some 600 homes in Dublin are currently at risk with 13,000 more potentially facing flooding if climate change continues at its present rate (Office of Public Works 2018a, b).

While the high cost of protecting high-value urban land from losses associated with sea-level rise can be justified, no such expenditure makes economic sense for agricultural and amenity lands. Planned retreat is the only feasible option here. In some cases, the ecosystem may be relocated naturally further inland. But the time to re-establish a salt marsh or machair environment is century-scale, as is the continuing rise in sea level. The loss of such environments, and the environmental goods and services they provide, is inevitable. Many of the salt marshes in Ireland, for example, provide valuable winter feeding grounds for many species of migratory birds and their loss would represent a further blow to biodiversity. In many cases, 'coastal squeeze' will occur where obstacles such as transport infrastructure or relief obstacles will result in extinction of the features concerned. Management of such areas has to be extremely sensitive since hard engineering solutions may have adverse consequences for adjacent areas. Wave modelling and good geological knowledge must be combined

**Table 1** Sea-level rise scenarios and cost in € millions for all insurance claims

| *Amount of sea-level rise* | *0.5 m* | *1 m* | *2 m* | *3 m* | *4 m* | *6 m* |
|---|---|---|---|---|---|---|
| Location | | | | | | |
| **Leinster** | | | | | | |
| Louth | 48 | 90 | 125 | 172 | 221 | 331 |
| Meath | 6 | 13 | 23 | 31 | 39 | 49 |
| Dublin | 170 | 339 | 513 | 678 | 898 | 1343 |
| Wicklow | 19 | 26 | 33 | 45 | 58 | 90 |
| Wexford | 13 | 25 | 40 | 62 | 83 | 131 |
| **Munster** | | | | | | |
| Waterford | 21 | 42 | 74 | 109 | 149 | 231 |
| Cork | 321 | 432 | 525 | 608 | 692 | 865 |
| Kerry | 13 | 21 | 39 | 73 | 133 | 255 |
| Limerick | 58 | 86 | 112 | 129 | 150 | 210 |
| Clare | 10 | 21 | 39 | 73 | 133 | 255 |
| **Connacht** | | | | | | |
| Galway | 14 | 37 | 63 | 98 | 138 | 265 |
| Mayo | 6 | 12 | 15 | 19 | 23 | 41 |
| Sligo | 5 | 13 | 20 | 28 | 35 | 56 |
| Leitrim | 0 | 0 | 0 | 0 | 0 | 1 |
| **Ulster** | | | | | | |
| Donegal | 21 | 42 | 57 | 79 | 97 | 154 |
| **Totals** | **725** | **1199** | **1673** | **2182** | **2792** | **4188** |

Source: Flood and Sweeney (2012)

with climatological expertise to ensure that wise long-term decisions are made. A Coastal Vulnerability Index is one way of approaching this and indicates where the hotspots needing priority consideration are located. One such index for a large part of eastern Ireland has recently been developed by Caloca Casado (2019).

The impacts of a rising sea level, combined with storm surge, are also dependent on when events happen in relation to the tide. Joint probability estimates are needed to estimate the extent to which coastal flooding will occur. The arrival of the storm surge and low-pressure system (which does itself elevate local sea level) and the extent to which a spring or neap tide is occurring are variables that need to be factored in. Thus far, the worst-case scenario of a high spring tide, full in when the storm arrives, has not been fully experienced. But ultimately the dice will fall unfavourably on this set of circumstances, and it must underpin safety arrangements for the populations at risk.

### *Sectors with Potential Short-Term Benefits*

It has long been known that extremes of temperature can provide an adverse impact on human health (Schneider and Breitner 2016). With high temperatures, the body employs several defence mechanisms to enhance heat losses, including dilation of the blood vessels, sweating, the production of anti-diuretic hormones and an increase in respiration. This places a requirement on the heart to work harder, and so many respiratory and cardiovascular diseases are exacerbated, resulting in increased mortality in vulnerable populations. By contrast, response to cold is usually mediated through a constriction of blood vessels in the skin and a diversion of blood supplies to the central core of the body in order to conserve heat. This also raises blood pressure and the heart rate and stresses cardiovascular functioning. Cold air also stresses the respiratory system by causing the smaller airways in the lung to constrict, and thus mortality in cold weather increases.

Many studies have described the relationship between temperature and mortality as a U-shaped curve, with mortality high at both hot and cold extremes and a minimum mortality band (Huynen et al. 2001). Using mortality data from Ireland which had been processed to remove issues arising from influenza epidemics, accidents, cancer and other causes of death not likely to be related to heat or cold, Cullen (2007) confirmed the existence of such a temperature-mortality relationship for Ireland. Indeed, a more recent study concluded that 'the island of Ireland currently has the highest levels of excess winter mortality in Europe, with an estimated 2,800 excess deaths during each winter... (The) key challenges are to develop and implement policies which tackle fuel poverty and reduce winter morbidity and mortality' (Zeka et al. 2014). Cullen's work indicated that minimum mortality occurred with mean daily temperatures somewhere between 15 and 19°C (Fig. 4). By contrast, sub-zero temperatures were associated with elevated values. At the other extreme, elevated values were also indicated for higher temperatures, but with less pronounced mortality increases than for cold. This confirms that a rise in mean daily temperatures will produce a fall in winter mortality from the causes described above, though less than the increase that might be expected from the same increase in summer heat. Of course, it is important to emphasise the caveats that are involved: the effect of improved health care, insulation and air conditioning, and so on, as well as contributions made by other heat-related health problems such as water-borne pathogens. But

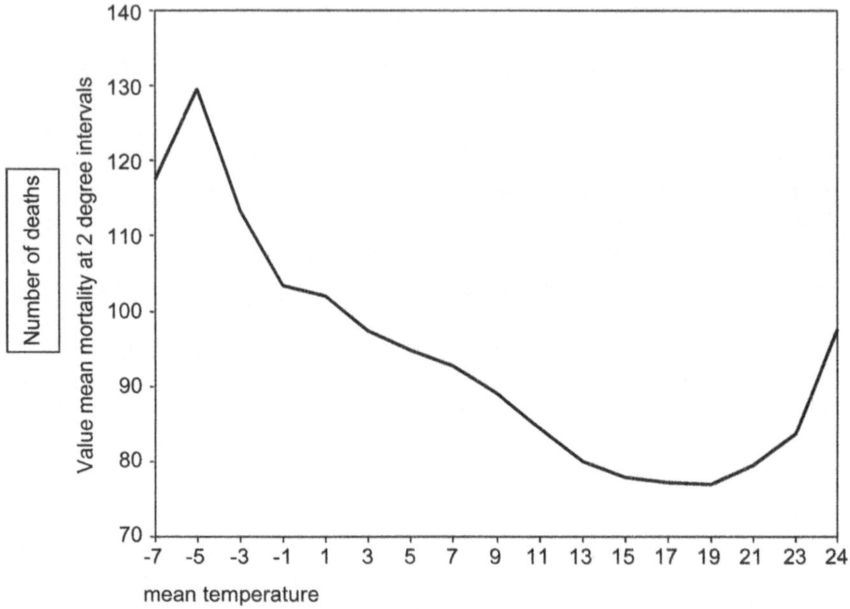


**Fig. 4** Mean mortality and temperature relationship for Ireland. (Source: Cullen 2007)

nonetheless, a modest resource shift in medical services towards the summer months may be in order as climate warms.

The other potential beneficiary of warming is the Irish tourist industry. Mieczkowski (1985) devised a Tourism Climate Index to express the climatic elements most affecting the quality of the international tourism experience and computed values of the index for 453 locations worldwide. This incorporated temperature, humidity, precipitation, sunshine and wind parameters into a weighted index where the score was interpreted as indicating comfort conditions for a typical tourist such as 'Extremely Unfavourable', 'Acceptable', 'Excellent' and so on. Salmon (2013) used Irish meteorological data from the present and modelled climate data for two future periods to estimate likely changes in tourism comfort in Ireland. It emerged that at present Irish tourism enjoys only three months (June, July, August) in the 'Good' category. However, after mid-century, these three months enter the 'Very Good' category. Secondly, an extension of the shoulder season is projected with May showing characteristics of the present June over the coming 30 years and September showing a similar comfort index after mid-century as present-day August. These are significant economic considerations for tourism development in Ireland. They also have to be viewed in terms of declining tourism comfort values over much of the Mediterranean Basin. In addition, a prognosis of increased deadly heatwaves in many subtropical cities exists, and potentially also later in the century in mid-latitude cities such as New York (Matthews et al. 2017).

## Climate Change and Ireland: Adaptation

Climate change will clearly create new vulnerabilities for Ireland in a number of sectors and heighten the risk of damages in existing ones. Mitigation on its own cannot address satisfactorily these impacts at a national scale since global emissions will determine their magnitude and extent. Adaptation involves planning for the protection of people, buildings, infrastructure, commercial activities and natural ecosystems to enable them to continue to function productively. Building resilience by incremental steps such as flood protection, tree planting or changing agricultural practices offers ways of adapting to the projected changes. In some cases, seeking to maintain existing systems may not be sufficient to cope with the rapidity of climate change, and more radical transformative adaptation mechanisms may be required involving a change in paradigm (IPCC 2012). Such an approach, however, may face socio-cultural, governance/institutional, resource-based and physical barriers (Clarke and Murphy 2019).

Traditional approaches to climate adaptation have emphasised a 'top-down' approach. This relied on climate change projections to be fed into impact models, the outputs of which would inform public decision-making processes. These would be designed to counter the changes in risk and provide greater resilience to the climate change hazard concerned. Making robust adaptation decisions on the basis of such a procedure has risks attached to it, however. These relate to the cascade of uncertainty that results from models being used to run other models. The tendency to interpret higher resolution downscaled scenarios as more accurate representations of future conditions also have to be tempered because of the inherent uncertainties transmitted. Although such top-down approaches have traditionally dominated adaptation approaches, decision makers have found it difficult to handle uncertainties without training and have often shied away from progressing the adaptation procedure beyond the impact assessment stage (Wilby and Dessai 2010).

A 'bottom-up' approach to adaptation has emerged in recent years. This emphasises consideration of what communities accept as representing successful coping with their known vulnerabilities. This can and should be informed by climate information on past, and likely future, climate extremes. Inevitably, this risk perception is shaped by multiple socio-economic factors and uncertainties regarding the extent to which future events may fall outwith past experience. But a 'no regrets' approach is not always the outcome.

Two Irish examples of where transformative adaptation has failed are described by Clarke and Murphy (2019). A bottom-up approach to addressing a flood problem was proposed by a local community group in Skibbereen, County Cork. This involved a multifunctional environmental park on the periphery of the town to provide a flood storage facility as well as a tourism and recreational resource. The local Council was instead envisaging using the area as a car park. Ultimately hard engineering solutions were decided on costing €14 m, and the bottom-up approach failed to progress. By contrast, in Clontarf County Dublin, a hard engineering solution proposed by Dublin City Council following a significant flood event in 2002 entailed a 3 km-long combination of flood wall/earthen bund ranging in height from 0.85 to 2.75 m along the promenade to offer protection for adjacent residential and commercial properties. Opposition to the loss of amenity this represented mobilised local community groups and gained political support sufficient to stall the project for ten years after planning permission was granted. In this case, perception of risk, valuation of social amenity and grassroots activism interacted to stall a top-down adaptation strategy. This is indicative of the complex interplay that adaptation to climate change in Ireland will likely face in the years ahead as it seeks to implement the National Adaptation Framework published in 2018.

## Climate Change and Ireland: Conclusions

Research capacity and capability in climate science has developed considerably in recent years, and Irish climate scientists continue to make important contributions to international bodies such as the IPCC. As the science and computer power has expanded, downscaled climate model development has also enabled greater clarity on what future climate scenarios

Ireland faces. Notwithstanding continuing uncertainty in detailed spatial and temporal resolutions, it is clear that the warming of 0.5°C which has occurred over the past 30 years is set to continue, reaching 1–1.5°C by mid-century under a 'business-as-usual' scenario. It will however be rainfall changes that prove to be most significant for Ireland, posing threats of increased winter flooding and summer droughts. These have implications for civil protection, agriculture and critical infrastructure which require detailed consideration and planning for, and significant expenditure to ensure preparedness and resilience is achieved in advance. Extremes are increasingly likely to herald the changes in Irish climate that will occur, and the attribution of a human-induced component to these is becoming increasingly evident.

Ireland's biodiversity will be susceptible to the climate changes projected, with particular habitats and species vulnerable to changed conditions and changed competition from invasive species and pests. Some losses of iconic landscapes such as peatlands and upland heaths will become evident and many bird, insect and fish species will have their ranges and distribution curtailed. Agriculture and forestry will not be immune to these changes either, with both grass growth and forest productivity especially vulnerable to drier summer conditions. But it is housing and commercial locations that will incur the greatest costs due to increased winter rain, especially where development has been allowed in vulnerable areas. The main urban centres will have additional risks imposed on them by increased sea level and storm surge activity, and coastal protection measures will be economically justified in such areas. Outwith these high-land-value areas, on the soft coastlines of rural Ireland, no rationale for hard engineered coastal protection will most likely be justified and a policy of coastal retreat may be the only feasible option. For intermediate areas, careful and sensitive management, involving detailed coastal vulnerability mapping, will be required.

Though there will be no long-term winners from climate change in Ireland, some sectors may gain some benefits. Tourism is likely to enjoy a longer and more reliable summer season, and health services are likely to see a diminution in cold-related winter mortality and morbidity. But these are in turn dependent on other factors such as air travel trends and waterborne pathogens, respectively, and should not be taken in isolation as good news stories.

Tensions, already evident with respect to adaptation, exemplify the major difficulties coming to terms with climate change are likely to pose for Ireland. Reconciling the demands and needs of different pressure groups and stakeholders will emerge as stumbling blocks in adaptation just as they have in mitigation. Faced with this, decision makers have difficulties in showing the decisive leadership necessary and will be tempted to respond more readily to short-term rather than long-term exigencies. Already identified internationally as a climate laggard, Ireland has responded to such internal pressures by not being identified as a country willing to show leadership in pushing for increased mitigation ambition within the EU. The emerging popular groundswell led by youth and non-governmental organisation (NGO) activists has the potential to change all of this. The social consensus which has characterised Irish responses to climate change so far has now shifted, and the issues have now moved beyond science and economics to a higher plane. Debate has commenced on what is an appropriate share of the remaining carbon budget that Ireland should seek to access (McMullin et al. 2019). As a 'nimble' country with a well-developed social conscience, Ireland, more than most countries, has the potential to provide an informed and ethically appropriate response to the challenge of climate change as well as ensuring a sustainable future for the next generation of its inhabitants.

## Bibliography

Byrne, C., M. Jones, and A. Donnelly. 2003. In *Climate Change: Scenarios and Impacts for Ireland*, ed. J. Sweeney et al., 229. Wexford: Environmental Protection Agency.

Caloca Casado, S. 2019. *Coastal Vulnerability Assessment of Co. Dublin and Co. Wicklow to Impacts of Sea-Level Rise*. Unpublished PhD thesis, Department of Geography, Maynooth University.

Carter, R.W.G. 1990. *The Impact on Ireland of Changes in Mean Sea Level*. Programme of Expert Studies Number 2. Dublin: Department of the Environment, 128pp.

Clarke, D., and C. Murphy. 2019. *Challenges of Transformative Climate Change Adaptation: Insight from Flood Risk Management*, Climate Change Research Programme (CCRP) 2007–2013, Report Series No. 265, 40. Wexford: Environmental Protection Agency.

Coll, J., and J. Sweeney. 2013. *Current and Future Vulnerabilities to Climate Change in Ireland*, Climate Change Research Programme (CCRP) 2007–2013, Report Series No. 29, 40. Wexford: Environmental Protection Agency.

Coll, J., D. Bourke, M. Sheehy Skeffington, M. Gormally, and J. Sweeney. 2014. Projected Loss of Active Blanket Bogs in Ireland. *Climate Research* 59: 103–115.

Coll, J., D. Bourke, R.L. Hodd, M. Sheehy Skeffington, M. Gormally, and J. Sweeney. 2016. Projected Climate Change Impacts on Upland Heaths in Ireland. *Climate Research* 69: 177–191.

Cox, P., M. Huntingford, and M. Williamson. 2018. Emergent Constraint on Equilibrium Climate Sensitivity from Global Temperature Variability. *Nature* 553: 319–322.

Cullen, E. 2007. *Climate Change and Health in Ireland: A National Vulnerability Assessment*. Unpublished PhD thesis, National University of Ireland Maynooth.

DELG. 2001. *Coastal Zone Management*. Dublin: Department of the Environment and Local Government, Spatial Planning Unit.

Dessai, S., and M. Hulme. 2004. Does Climate Adaptation Policy Need Probabilities? *Climate Policy* 4: 107–128. https://doi.org/10.1080/14693062.2004.9685515.

Donnelly, A., A. Cafffara, F. Albanito, and M. Jones. 2008. In *Climate Change in Ireland: Refining the Impacts*, ed. J. Sweeney et al., 163. Wexford: Environmental Protection Agency.

Fealy, R., and J. Sweeney. 2008. Statistical Downscaling of Temperature, Radiation and Potential Evapotranspiration to Produce a Multiple GCM Ensemble Mean for a Selection of Sites in Ireland. *Irish Geography* 41 (1): 1–27.

Fealy, R., C. Bruyére, and C. Duffy. 2018. *Regional Climate Model Simulations for Ireland for the 21st Century.*, EPA Research Report. Vol. 244, 137. Wexford: Environmental Protection Agency.

Flood, S., and J. Sweeney. 2012. Quantifying Impacts of Potential Sea-Level Rise Scenarios on Irish Coastal Cities. In *Resilient Cities 2011*, ed. K. Otto-Zimmermann, 27–52. London: Springer.

Gleeson, E., R. McGrath, and M. Treanor, eds. 2013. *Ireland's Climate: The Road Ahead*. Dublin: Met Éireann.

Holden, N., A. Brereton, R. Fealy, and J. Sweeney. 2003. Possible Change in Irish Climate and Its Impact on Barley and Potato Yields. *Agricultural and Forest Meteorology* 116: 181–196.

Huynen, M.M., P. Martens, D. Schram, M.P. Weijenberg, and A.E. Kunst. 2001. The Impact of Heat Waves and Cold Spells on Mortality Rates In the Dutch Population. *Environmental Health Perspectives* 109 (5): 463–470. https://doi.org/10.1289/ehp.01109463.

IPCC. 2012. Managing the Risks of Extreme Events and Disasters to Advance Climate Change Adaptation. In *Managing the Risks of Extreme Events and Disasters to Advance Climate Change Adaptation: Special Report of the Intergovernmental Panel on Climate Change*, ed. C.B. Field, V.R. Barros, T. Stocker, D. Qin, D. Dokken, K.L. Ebi, M. Mastrandrea, K.J. Mach, G. Plattner, S. Allen, M. Tignor, and P. Midgely. Cambridge: Cambridge University Press.

———. 2018. Summary for Policymakers. In *Global Warming of 1.5°C. An IPCC Special Report on the Impacts of Global Warming of 1.5°C Above Pre-industrial Levels and Related Global Greenhouse Gas Emission Pathways, in the Context of Strengthening the Global Response to the Threat of Climate Change, Sustainable Development, and Efforts to Eradicate Poverty*, ed. V. Masson-Delmotte, P. Zhai, H.-O. Pörtner, D. Roberts, J. Skea, P.R. Shukla, A. Pirani, W. Moufouma-Okia, C. Péan, R. Pidcock, S. Connors, J.B.R. Matthews, Y. Chen, X. Zhou, M.I. Gomis, E. Lonnoy, T. Maycock, M. Tignor, and T. Waterfield. Geneva: World Meteorological Organization, 32 pp.

———. 2019a. *Summary for Policymakers*, Special Report on the Ocean and Cryosphere in a Changing Climate. Geneva: World Meteorological Organization.

———. 2019b. *Summary for Policymakers. Climate Change and Land: an IPCC special Report on Climate Change, Desertification, Land Degradation, Sustainable Land Management, Food Security, and Greenhouse Gas Fluxes in Terrestrial Ecosystems.* Geneva: World Meteorological Organization.

———. 2019c. *2019 Refinement to the 2006 IPCC Guidelines for National Greenhouse Gas Inventories.* Geneva: World Meteorological Organization.

Jackson, D.W.T., and J.A.G. Cooper. 2011. Coastal Dune Fields in Ireland: Rapid Regional Response to Climatic Change. *Journal of Coastal Research*, SI 64 (Proceedings of the 11th International Coastal Symposium), Szczecin, ISSN 0749-0208.

Levin, K., B. Cashore, S. Bernstein, and G. Auld. 2012. Overcoming the Tragedy of Super Wicked Problems: Constraining Our Future Selves to Ameliorate Global Climate Change. *Policy Sciences* 45: 123–152. https://doi.org/10.1007/s11077-012-9151-0.

Matthews, T., C. Murphy, R. Wilby, and S. Harrigan. 2014. Stormiest Winter on Record for Ireland and UK. *Nature Climate Change* 4: 738–740.

Matthews, T., R. Wilby, and C. Murphy. 2017. Communicating the Deadly Consequences of Global Warming for Human Heat Stress. *Proceedings of the National Academy of Sciences* 114 (15): 3861–3866. https://doi.org/10.1073/pnas.1617526114.

McGrath, R., and P. Lynch, eds. 2008. *Ireland in a Warmer World*, 109. Dublin: Community Climate Change Consortium for Ireland (C4I).

McGrath, R., E. Nishimura, P. Nolan, T. Semmler, C. Sweeney, and S. Wang. 2005. *Climate Change: Regional Climate Model Predictions for Ireland.* Environmental Protection Agency, ERTDI Report Series No. 36, 45.

McMullin, B., P. Price, M.B. Jones, and A.H. McGeever. 2019. Assessing Negative Carbon Dioxide Emissions from the Perspective of a National "Fair Share" of the Remaining Global Carbon Budget. *Mitigation and Adaptation Strategies for Global Change*. https://doi.org/10.1007/s11027-019-09881-6.

Mieczkowski, Z. 1985. The Tourism Climatic Index: A Method of Evaluating World Climates for Tourism. *The Canadian Geographer* 29: 220–233. https://doi.org/10.1111/j.1541-0064.1985.tb00365.x.

Murphy, C., C. Broderick, T. Burt, M. Curley, C. Duffy, J. Hall, S. Harrigan, T. Matthews, N. Macdonald, G. McCarthy, M. McCarthy, D. Mullan, S. Noone, T. Osborn, C. Ryan, J. Sweeney, P. Thorne, S. Walsh, and R. Wilby. 2018. A 305-Year Continuous Monthly Rainfall Series for the Island of Ireland (1711–2016). *Climate of the Past* 14: 413–440.

Nolan, P. 2015. *Ensemble of Regional Climate Model Projections for Ireland*, EPA Climate Change Research Report No. 159. Wexford: EPA.

NPWS. 2013. *The Status of Protected EU Habitats and Species in Ireland*. Overview Volume 1. Unpublished Report, National Parks & Wildlife Services, Department of Arts, Heritage and the Gaeltacht, Dublin.

Office of Public Works. 2018a. *Implementing the National Flood Risk Policy*. Meath: Office of Public Works, Trim, Co.

———. 2018b. *Flood Risk Management Plan: River Basin (09) Liffey & Dublin Bay*. Meath: Office of Public Works, Trim, Co.

Rahmstorf, S. 2008. Anthropogenic Climate Change: Revisiting the Facts. In *Global Warming: Looking Beyond Kyoto*, ed. E. Zedillo, 34–53. Washington: Brookings Institution Press.

Ray, D., G. Xenakis, T. Semmler, and K. Black. 2007. The Impact of Climate Change on Forest in Ireland and Some Options for Adaptation. In *Proceedings of the COFORD Conference: Forests, Carbon and Climate Change*, ed. Eugene Hendrick and Kevin Black, 27–33. Wicklow: Glenview Co.

Rittel, H., and M. Webber. 1973. Dilemmas in a General Theory of Planning. *Policy Sciences* 4 (2): 155–169. https://doi.org/10.1007/BF01405730.

Salmon, N. 2013. Tourism and Climate Change in Ireland – Impacts and Adaptation. In *Co-ordination, Communication and Adaptation for Climate Change in Ireland: An Integrated Approach (COCOADAPT)*, EPA Climate Change Research Programme (CCRP) 2007–2013 Report Series No. 30, ed. J. Sweeney, D. Bourke, J. Coll, S. Flood, M. Gormally, J. Hall, J. McGloughlin, C. Murphy, N. Salmon, M. Sheehy Skeffington, and D. Smyth. Wexford: Environmental Protection Agency, 118pp.

Satoh, M., A. Noda, T. Seiki, Y. Chen, C. Kodama, Y. Yamada, N. Kuba, and Y. Sato. 2018. Toward Reduction of the Uncertainties in Climate Sensitivity Due to Cloud Processes Using a Global Non-Hydrostatic Atmospheric Model. *Progress in Earth and Planetary Science* 5: 67.

Schneider, A., and S. Breitner. 2016. Temperature Effects on Health – Current Findings and Future Implications. *eBioMedicine* 6: 29–30. https://doi.org/10.1016/j.ebiom.2016.04.003.

Sweeney, J., T. Brereton, C. Byrne, R. Charlton, C. Emblow, R. Fealy, N. Holden, M. Jones, A. Donnelly, S. Moore, P. Purser, K. Byrne, E. Farrell, E. Mayes, D. Minchin, J. Wilson, and J. Wilson. 2003. *Climate Change: Scenarios & Impacts for Ireland*. Wexford: Environmental Protection Agency.

Sweeney, J., F. Albanito, A. Brereton, A. Caffarra, R. Charlton, A. Donnelly, R. Fealy, J. Fitzgerald, N. Holden, M. Jones, and C. Murphy. 2008. *Climate Change: Refining the Impacts for Ireland*. Wexford: Environmental Protection Agency.

Thackeray, C., and C. Fletcher. 2016. Snow Albedo Feedback: Current Knowledge, Importance, Outstanding Issues and Future Directions. *Progress in Physical Geography: Earth and Environment* 40 (3): 392–408.

Wilby, R., and S. Dessai. 2010. Robust Adaptation to Climate Change. *Weather* 65 (7): 180–185.

Williams, C., R. Fealy, R. Teck, J. Coll, J. Rigby, J. Sweeney, and C. Griffin. in prep. Validation of a Voltinism Model of Hylobius Abietis (L.) with Predictions for Ireland Based on Climate Change. *Agriculture and Forest Entomology*.

World Weather Attribution. 2019. *Human Contribution to the Record-Breaking July 2019 Heatwave in Western Europe – World Weather Attribution*. https://www.worldweatherattribution.org/human-contribution-to-the-record-breaking-july-2019-heat-wave-in-western-europe/World Weather Attribution. Retrieved July 19, 2020.

Zeka, A., S. Browne, H. McAvoy, and P. Goodman. 2014. The Association of Cold Weather and All-Cause and Cause-Specific Mortality in the Island of Ireland Between 1984 and 2007. *Environmental Health* 13 (1): 104–112. ISSN: 1476-069X.

# Ireland's Policy Response to Climate Change: An Historical Overview

*Diarmuid Torney*

## Introduction

Ireland's policy response to climate change stands at a crossroad. On the one hand, the state has been regarded for some time as a climate laggard, repeatedly ranked among the lowest performing European Union (EU) member states in the annual Climate Change Performance Index (Burck et al. 2017, 2018) and on track to miss EU climate targets by a considerable margin (EPA 2019a). Although there has been progress in some areas, most notably deployment of renewable electricity, Ireland's progress in responding to climate change has been hindered by a failure to tackle emissions from key sectors including agriculture and transport. On the other hand, the period since late 2017 has witnessed a wave of institutional and policy innovations, including surprisingly far-reaching recommendations by the Citizens' Assembly on climate change published in March 2018 (Citizens' Assembly 2018), an influential and, in many

D. Torney (✉)
School of Law and Government, Dublin City University, Dublin, Ireland
e-mail: diarmuid.torney@dcu.ie



D. Robbins et al. (eds.), *Ireland and the Climate Crisis*, Palgrave Studies in Media and Environmental Communication,
https://doi.org/10.1007/978-3-030-47587-1_3

respects, radical report published in March 2019 by a parliamentary committee established to consider the Assembly's recommendations (Houses of the Oireachtas 2019), and a landmark Climate Action Plan published in June 2019 (DCCAE 2019).

A limited amount of academic and policy research has focused on Ireland's response to climate change, though a broader body of research has examined environmental politics in Ireland (Barry 2009; Curtis and Lyons 2011; Flynn 2003, 2006; Laffan and O'Mahony 2008; O'Mahony and Keohane 2011; Taylor 1998; Taylor and Horan 2001). Earlier research on Ireland's response to climate change focused on how international and domestic factors shaped Ireland's response to climate change (Coghlan 2007), cross-cutting studies of policy choices facing Ireland (Brennan and Curtin 2008; NESC 2012), analyses of the institutional context shaping Ireland's response to climate change (Kirby 2013; McGloughlin and Sweeney 2011; Ó'Broin and Kirby 2015), studies of the adoption of a carbon tax (Convery et al. 2013; Harrison 2010), as well as analyses of extreme weather events and projections of future climate impacts (Cunningham 2008; Hickey 2008, 2010).

More recently, the literature has expanded to include studies of the development of Ireland's 2015 climate law (Torney 2017, 2019), climate change and party politics (Ladrech and Little 2019; Little 2017), Ireland's climate change governance (NESC 2019), and the impact of EU membership on Ireland's climate policy response (Torney and O'Gorman 2019). Meanwhile, sectoral studies have drawn attention to the opportunities for, and constraints on, low-carbon transition in energy (Torney 2018), agriculture (Curtin and Arnold 2016), and transport (Devaney and Torney 2019).

This chapter provides an overview of Ireland's policy response as well as the drivers and constraints on climate action. The chapter traces the evolution of climate policy during four distinct historical phases. The first of these is the era of the so-called Celtic Tiger, which saw a rapid rise in emissions. The second was characterised by profound crisis of economy and polity, during which attention was diverted from climate policy but with a significant fall in greenhouse gas (GHG) emissions. The third phase is the economic recovery, which in many ways was as unexpected in its speed and extent as the crash. During this period, Ireland's emissions rebounded strongly, illustrating that Ireland's emissions remain stubbornly coupled to economic growth. Fourth and finally, the chapter examines Irish climate policy since 2017, noting a range of innovative and progressive developments including the Citizens' Assembly, the Joint Oireachtas Committee on Climate Action, and the 2019 Climate Action Plan. This section addresses the question of whether Ireland has left its laggard status behind. Finally, the chapter reflects on the broader context that has shaped Ireland's policy response to climate change, including economic structure and political and governance frameworks.

## The Celtic Tiger

Ireland's economic development occurred late by comparative European standards. A relatively poor economy for much of the period since independence in the 1920s, Ireland caught up with European average standards of living only in the 1990s (Barry et al. 1999). During the 1990s and early- to mid-2000s, Ireland's so-called Celtic Tiger economy grew by double digit growth rates per annum during the period 1995–2000 and by high single figure growth rates per annum between 2000 and 2007 (CSO 2020). Under the Kyoto Protocol, the EU faced a GHG reduction target of 8 percent overall. Differentiated targets were given to the EU's then 15 member states. Under this so-called burden-sharing agreement, Ireland was allocated a domestic target of limiting emissions growth to not more than 13 percent above 1990 levels. This target was designed to take account of the low starting point of the Irish economy and the catch-up Ireland was undertaking over the period in question relative to its European counterparts.

The profile of Ireland's GHG emissions was strongly upwards during the 1990s. Emissions rose by nearly a quarter over the decade of the 1990s and peaked in 2001 at 27 percent above 1990 levels. Over the period 2001–2007, emissions were essentially flat, with a slight decrease of 3.5 percent over the period (EPA 2019b). The principal driver of emissions growth was the rapid increase in economic activity. A reduction in emissions intensity driven by efficiency gains accounts for the essentially flat emissions trajectory over the 2000s (O'Mahony 2013). Nonetheless, by 2007 Ireland's emissions stood at 23 percent above 1990 levels, not far short of double Ireland's Kyoto Protocol target of an increase of 13 percent.

In terms of domestic climate policy, the early- to mid-2000s were marked principally by the publication of two overarching climate change policy documents. The first of these, Ireland's first National Climate Change Strategy, was published in 2000 (DELG 2000). It outlined a range of measures intended to enable the state to meet Ireland's target under the EU burden-sharing agreement to limit emissions growth to no more 13 percent above 1990 levels over the period 2008–2012.

Shortly before leaving office, the 2002–2007 Fianna Fáil-Progressive Democrats government published a second National Climate Change Strategy in April 2007 (DEHLG 2007). This aimed to set a pathway towards meeting Ireland's EU emissions reduction target. Notably, this included significant recourse to purchasing of credits under the UN's so-called Clean Development Mechanism, which entails financial transfers to developing countries to support emissions reductions instead of undertaking emissions reductions domestically. Another noteworthy policy document from this era was the Energy Policy Framework 2007–2020, *Delivering a Sustainable Energy Future for Ireland*. It framed Ireland's response to energy challenges in terms of broader European and global trends and approaches, noting a "global and European context which has put energy security and climate change among the most urgent international challenges" (DCMNR 2007: 5).

The 1990s and 2000s saw the beginnings of a policy response to climate change. In the electricity sector, the first support schemes for renewable electricity in Ireland were developed during this period. The first of these, the Alternative Energy Requirement scheme, introduced in 1996, provided winning bidders with a 15-year power purchase agreement whereby the Electricity Supply Board (ESB) bought the electricity output of the winning facility at the bid price. This was followed by the REFIT 1 scheme, which opened in 2005 and provided long-term feed-in tariffs to encourage new capacity development of proven technologies and short-term feed-in tariffs to support the development of emerging ocean energy technologies. However, to a considerable extent, climate policy development in Ireland remained limited during the Celtic Tiger era.

## Economic Crisis

Following a general election in June 2007, the Programme for Government of the incoming coalition government between Fianna Fáil, the Green Party and the Progressive Democrats included a pledge to reduce GHG emissions by 3 percent per year. However, negative economic circumstances soon intervened. Ireland was hit by profound economic crisis in the late 2000s, driven by a collapse of the property sector that precipitated a severe banking crisis. During the period 2007–2009, Irish Gross National

Income fell by over 11 percent. In November 2010, the Irish government was compelled to request international assistance from the EU and the International Monetary Fund in the form of a €85 billion bailout. While exacerbated by the broader global financial crisis, Ireland's economic and banking crisis was primarily home-grown, brought about by inadequate risk management practices in Irish banks and a failure of regulatory oversight (O'Sullivan and Kennedy 2010).

The crisis had a pronounced effect on Ireland's GHG emissions, which fell by 15 percent between 2007 and 2011, and by 8.5 percent in a single year, from 2008 to 2009 (EPA 2019b). This decline was driven for the most part by the precipitous fall in economic activity rather than by climate change mitigation policies. Nonetheless, the result was that the state managed, largely by accident, to achieve compliance with its Kyoto Protocol first commitment period (2008–2012) target to limit growth in emissions to not more than 13 percent above 1990 levels (EPA 2013).

In many respects, climate change disappeared from the policy agenda during the crisis years. Nonetheless, this period saw the elaboration of the EU's climate and energy package for 2020, with corresponding obligations for Ireland. These included a reduction of GHG emissions in sectors outside of the Emissions Trading Scheme (ETS) of 20 percent relative to 2005 levels, and an overall renewable energy target of 16 percent, composed of a 40 percent target for electricity, a 12 percent target for heat, and a 10 percent target for transport (Government of Ireland 2010; see also chapter "Climate Law in Ireland: EU and National Dimensions" by O'Gorman, this volume).

A carbon tax was introduced in 2010, set initially at €15 per tonne on liquid fossil fuels outside the ETS and increased to €20 per tonne in 2012. It was extended to solid fossil fuels in 2013 at €10 per tonne, which was increased to €20 per tonne in 2014. Under the Fianna Fáil/Green Party government, a "carbon budget" was introduced, but not of the sort provided for in the UK Climate Change Act. Rather, this was a statement by the Minister for the Environment following the finance minister on each Budget Day on the state's progress towards meeting emissions targets and actions undertaken.

The period also saw an unsuccessful attempt by Green Party environment minister John Gormley to introduce framework climate change legislation (Torney 2017). Gormley published an outline of the proposed legislation in December 2009 and draft legislation in December 2010. The Climate Change Response Bill would have legislated for annual average reduction targets of 2.5 percent for the period from 2008 to 2020 as well as a 40 percent reduction target for 2030 and an 80 percent reduction target for 2050. However, in January 2011, the Green Party announced its decision to withdraw from government, and the legislation was never enacted.

Sectoral policy development and implementation was largely constrained also by the prevailing economic circumstances during this period. Nonetheless, there were some positive developments. The electricity sector saw significant deployment of renewable generation capacity, principally wind. This was underpinned by additional state support schemes. In 2012, the REFIT 2 and REFIT 3 schemes were launched. REFIT 2 provided feed-in tariff support for small and large onshore wind, small hydro, and biomass/landfill gas. The REFIT 3 scheme was designed to incentivise high efficiency combined heat and power, biomass combustion, and biomass co-firing, though biomass combustion and biomass co-firing with peat would become the subject of scrutiny and criticism on environmental grounds.

In the transport sector, *Smarter Travel—A Sustainable Transport Future* was published in 2009 and committed to reducing overall travel demand, maximising efficiency of the transport network, reducing reliance on fossil fuels, reducing transport emissions, and improving transport accessibility (DTTAS 2009). It was, in many ways, a visionary document, but the economic crisis, which was not fully foreseen at the time of publication of the strategy, severely constrained investment in the transport sector in the period since 2009, and the targets set out in *Smarter Travel* were for the most part not met.

## Return to Growth

Just as unexpected as the severity of the economic crash was the speed and strength of the recovery from the early 2010s onwards, with Ireland registering the fastest economic growth rate in the EU for several years running. Greenhouse gas emissions rebounded broadly in line with economic recovery, with an increase of over 6 percent between 2011 and 2018, at a time when emissions were supposed to be declining in accordance with the EU 2020 climate and energy framework (EPA 2019b). However, this increase was uneven, with most of it occurring in the years 2014–2016. Moreover, the agriculture and transport sectors were the big drivers of the growth, both rising 9.6 percent during the period 2011–2018.

The government elected in 2011 and composed of Fine Gael and Labour committed in its programme for government to introduce framework climate legislation. However, it was not until December 2015 that the Climate Action and Low Carbon Development Act was enacted. This was preceded by agreement by cabinet in 2014 of a "National Policy Position on Climate Action and Low Carbon Development". This latter document committed to an aggregate 80 percent reduction of emissions across the electricity generation, built environment, and transport sectors by 2050 and "an approach to carbon neutrality in the agriculture and land-use sector, including forestry, which does not compromise capacity for sustainable food production" (Government of Ireland 2014). The climate law passed in 2015 established a planning and reporting framework for climate change policy (see chapter "Climate Law in Ireland: EU and National Dimensions" by O'Gorman, this volume for further discussion). Most notably in an international context, the law did not set a quantified mitigation target even for 2050, making Ireland's law unusual by international standards (Ecologic 2020).

The 2015 climate law also established an independent Climate Change Advisory Council with a remit to provide advice to government and to assess progress in achieving national policy goals. The Council has not shied away from critical commentary on the government's response to climate change. For example, the Council's 2018 Annual Review noted that "Ireland is completely off course in terms of achieving its 2020 and 2030 emissions reduction targets. Without urgent action that leads to tangible and substantial reductions in greenhouse gas emissions, Ireland is unlikely to deliver on national, EU and international obligations" (CCAC 2018: iii). The Council's 2019 Annual Review struck a more positive note, welcoming publication of the government's Climate Action Plan, though it expressed concern over the "continued failure to set out detailed pathways on the cost-effective route to decarbonising the Irish economy by 2050" (CCAC 2019: iv).

The first National Mitigation Plan (NMP), the publication of which was required under the climate law, was published by the government in July 2017 (DCCAE 2017). It detailed a range of existing and planned measures across the electricity generation, built environment, and transport sectors to reduce GHG emissions. When publishing the Plan, the government described it as a "living document" and acknowledged that it "does not provide a complete roadmap to achieve the 2050 objective, but begins the process of development of medium to long term mitigation choices for the next and future decades" (DCCAE 2017). In a landmark judgement delivered in July 2020, the Supreme Court quashed the NMP. In a case taken by environmental NGO Friends of the Irish Environment, the court ruled that the NMP failed to comply with the requirements of the 2015 Climate Action and Low Carbon Development Act (see chapter "Climate Litigation, Politics and Policy Change: Lessons from Urgenda and Climate Case Ireland" by O Neill and Alblas, this volume).

The deployment of renewable energy and related infrastructure, particularly wind but also grid development, has faced increasing opposition from communities (NESC 2014). This led to many proposed projects being challenged through the courts. Nonetheless, an Energy White Paper published in 2015 committed to transforming Ireland's energy sector into a clean, low-carbon system by 2050, stating that "eventually, we will have to generate 100 percent of all our energy needs—not just electricity—from clean sources" (DCENR 2015).

In the agriculture sector, GHG emissions rose rapidly during the 2010s. Although only marginally above 1990 levels in 2018, GHG emissions from the agriculture sector increased by 15 percent between 2011 and 2018 (EPA 2019b). This was driven significantly by a substantial expansion of the beef and dairy sectors. This has been underpinned to a large extent by national policy signals, including the Food Harvest 2020 and Food Wise 2025 government strategies. These prioritised expansion of beef and milk exports in particular. Furthermore, the abolition of EU milk quotas in 2015 was another significant contributory factor.

Adaptation policy, long the poor relation of mitigation, gained some traction during the 2010s. In 2012, a National Climate Change Adaptation Framework was published by government to provide a framework for national and sectoral adaptation development (DECLG 2012). While it was non-statutory, it led to the development of sectoral adaptation plans for the agriculture and transport sectors developed. The 2015 climate law put adaptation policy on a firmer statutory basis. A first statutory National Adaptation Framework was published by government under the climate law in February 2018 (DCCAE 2018b). This was followed by guidelines developed by the Department of Communications, Climate Action and Energy for the development of sectoral and local authority adaptation plans (DCCAE 2018a, c).

## A New Dawn?

The period since 2017 has witnessed a new momentum in Ireland's response to climate change. An important driver of this was the process and outcomes of the Citizens' Assembly of 2016–2018, which considered, among other topics, the question of "How the state can make Ireland a leader in tackling climate change" over two weekends in autumn 2017. Interestingly, climate change was not specified as a topic for the Assembly to consider in the draft resolution introduced in the Oireachtas in July 2016. However, as a result of a Green Party amendment, climate change was added to the agenda.

An exceptional experiment in democratic governance, the Citizens' Assembly comprised 99 citizens drawn from all walks of life and afforded them the time, space, and structure to consider complex questions of public policy in a deliberative way. Climate change was one of five topics considered by the Assembly. The recommendations they agreed on the climate change topic were significantly more radical than many observers expected, with strong support (80 percent or higher in all cases) for all proposed actions, including paying higher taxes on carbon-intensive activities, taxing GHG emissions from agricultural produce, and phasing out support for peat extraction on a phased basis (Citizens' Assembly 2018).

In the period following on from the conclusion of the Citizens' Assembly's deliberations on climate change, a number of other factors contributed to raising the profile of climate change on the national and international policy agenda. First, a landmark report on *Global Warming of 1.5°* published by the Intergovernmental Panel on Climate Change (IPCC 2018) painted in stark terms the consequences of failing to address climate change and attracted unprecedented media and public attention. Second, on the day following the publication of the IPCC report, then Minister for Finance Pascal Donohoe deferred a highly anticipated decision to increase the rate of the carbon tax, generating significant negative commentary. Third, the school strikes movement, inspired by Swedish teenager Greta Thunberg, brought unprecedented numbers of people to the streets to protest for stronger government action on climate change (see chapter "The Changing Faces of the Climate Movement in Ireland" by Gold, this volume).

An all-party Joint Oireachtas Committee on Climate Action (JOCCA) was set up in July 2018 to consider the Citizens' Assembly's recommendations on climate change. Its report, published in March 2019, set out over

40 priority recommendations that were similarly far-reaching and endorsed many of the citizens' proposals (Houses of the Oireachtas 2019). Significant media attention focused on political disagreements over a proposed increase in the carbon tax. Fine Gael, Fianna Fáil, Labour, and the Green Party supported increasing the carbon tax, while Sinn Féin and People Before Profit opposed it.

Notwithstanding the very public political disagreement on carbon tax, there was cross-support for most of the JOCCA's recommendations (see also chapter "The Party Politics of Climate Change in Ireland" by Little, this volume). These included a range of measures to deal with GHG emissions from the key sectors of agriculture, energy, transport, and buildings, as well as recommendations for a just transition, the role of citizens and communities, and provisions around climate change education and communication. Among the most important elements of the JOCCA report were its recommendations for a new framework for governing the response to climate change (see also chapter "Climate Law in Ireland: EU and National Dimensions" by O'Gorman, this volume).

In turn, the JOCCA report played a significant role in shaping an all-of-government *Climate Action Plan to Tackle Climate Breakdown*, published by Minister Richard Bruton in June 2019 (DCCAE 2019). Although some of the sectoral chapters—particularly those on agriculture and transport—could have been more ambitious, the provisions on governance mirrored closely the recommendations of the JOCCA report. Moreover, they were modelled closely on the 2008 UK Climate Change Act. This is particularly noteworthy given that the UK model was considered but rejected during the framing of the 2015 climate change law because of objections by political parties and interest groups (Torney 2017). On 8 January 2020, the government published draft heads of the Climate Action (Amendment) Bill (DCCAE 2020), but two days later Taoiseach Leo Varadkar called a general election and the draft did not proceed further through the legislative process. A new government, formed in June 2020 and comprising Fianna Fáil, Fine Gael and the Green Party, committed to carrying forward this process by enacting a Climate Action (Amendment) Bill. It also committed to an average annual decarbonisation rate of 7% over the course of the decade 2021–2030 (Government of Ireland 2020).

## The Context Shaping Ireland's Response to Climate Change

It is clear that Ireland's policy response to climate change has been insufficient to date, though the period since 2017 has witnessed new momentum. As a small European state, Ireland faces conflicting pressures with respect to climate change (Carter et al. 2019; Little and Torney 2017). Ireland's policy response to climate change has been shaped by a range of political, institutional, societal, and economic factors.

Politically, as Conor Little argues cogently in his contribution to this volume, climate change has been a low-priority issue for most Irish political parties. There have been exceptions, most notably the periods 2006–2007 and since 2019, both of which coincided with strong economic growth and low unemployment. Institutionally, Ireland is heavily centralised, with limited powers devolved to local authorities (see chapter "Responding to Climate Change: The Role of Local Government in Ireland" by Dekker, this volume). This can inhibit experimentation and the development of bespoke policy responses for regional and local circumstances, which can be important in developing responses to climate change (Devaney and Torney 2019). Attempts at community and societal engagement remain sporadic, notwithstanding important exceptions such as the Citizens' Assembly and some of the initiatives discussed by Clare Watson and Peadar Kirby in their contributions to this volume.

In another respect, however, Ireland's governance landscape is fragmented horizontally. Recent decades have witnessed a progressive "agencification" of the Irish state. Comparative research has found that, across sectors, regulatory agencies overall in Ireland enjoy the highest level of formal independence across 17 West European countries (Gilardi 2005). Moreover, agencies have been created in Ireland in a relatively ad hoc fashion, with a wide variety of accountability and reporting relationships with central government (MacCarthaigh 2011).

Where powers have been delegated to independent regulators, this effectively introduces new veto players over which central government has relatively little control (Lockwood et al. 2017). This is indeed the case in respect of both the energy and transport sectors in Ireland (Devaney and Torney 2019; Torney 2018). Such fragmented decision-making structures make the task of decarbonisation of key sectors of the economy all the harder.

The possibilities for, and constraints on, Ireland's policy response to climate change have also been shaped in important ways by the structure of the economy, both in the present and historically. Ireland's economy is dominated by services, which accounted for nearly 60 percent of gross value added in 2018. Industry accounted for 37 percent, while the primary sector accounted for just 1 percent of gross value added (OECD 2020). This is striking for a country the economy of which was historically dominated by agriculture. However, the broader agri-food sector accounted for 7 percent of gross value added in 2016 (Teagasc 2020). Nonetheless, the historical legacy of a predominantly agricultural society gives the farming sector a voice in national policymaking that arguably is disproportionate to its contemporary contribution to the national economy.

Ireland's economic history matters for climate policy in other ways too. Because of its late industrialisation, Ireland does not have the same legacy of heavy industry that has characterised many other industrialised countries. As a result, Ireland's emissions profile has not benefited to the same extent from trends towards de-industrialisation resulting in "offshoring" of heavy industry, which benefits the national emissions inventories of countries outsourcing heavy industry because of the production-based methods of carbon accounting.

Ireland's dispersed pattern of spatial development and comparatively low population density by comparison with other industrialised countries also shapes the context for climate policy development. At 70 persons per square kilometre, Ireland's population density is considerably below the EU28 average of 117.7 persons per square kilometre (Eurostat 2020). This includes significant development of "one-off housing"—houses built in rural areas in recent decades that are not in close proximity to existing houses. This reduces the viability of public transport and increases car dependency. A more dispersed rural population is one factor underpinning objections to wind farm developments, which have become increasingly prevalent in recent years. In February 2018, the government unveiled a new National Planning Framework with a 2040 time horizon as well as a ten-year capital expenditure programme. Combined under the heading "Project Ireland 2040", it aimed to encourage the development of denser urban development and to limit suburban sprawl.

The broader European context has also been important, though it should not be overstated (see chapter "Climate Law in Ireland: EU and National Dimensions" by O'Gorman, this volume for a detailed discussion). The Irish state has formally complied with a range of planning and reporting obligations under EU legislation, such as the elaboration of National Renewable Energy Action Plans and National Energy Efficiency Action Plans and periodic reporting of GHG emissions inventories and projections as well as climate change policies and measures. However, if we judge the impact of the EU's impact in terms of delivery on Ireland's substantive as opposed to procedural obligations, the picture is much more mixed. Indeed, perhaps what is most striking is how limited the impact of EU membership has been on constraining Ireland's policy substantive response to climate change (Torney and O'Gorman 2019).

## Conclusion

Long regarded a climate laggard, Ireland entered the third decade of the twenty-first century with surprising momentum. This is not before time. Progress has been made in decarbonising electricity generation, but reducing GHG emissions from other sectors has proved much more challenging. Reduction in emissions from electricity sector has been driven by deployment of renewables and fuel switching, and the 2019 Climate Action Plan committed to a target of 70 percent renewable electricity by 2030. The state has hardly begun to tackle the agriculture and transport sectors, and the residential and industrial sectors pose significant challenges as well.

Impossible to predict just a few short years ago, the chain of events leading from the Citizens' Assembly through the Joint Oireachtas Committee on Climate Action to the 2019 Climate Action Plan transformed Ireland's climate policy landscape. While some of the sectoral chapters of the Climate Action Plan could have been more ambitious, the governance provisions of the Plan in particular have the potential to be transformational. This will be required if the state is to meet even its existing commitments under the EU's 2030 climate and energy framework. These commitments could be strengthened further in the context of the European Commission's *European Green Deal* proposal published in December 2019 (European Commission 2019). What impact the Covid-19 pandemic will have on Irish, European and global climate action remains uncertain (see introduction to this volume for initial reflections on this theme).

The February 2020 general election upended Irish politics. For the first time in any election since 1918, Sinn Féin received the largest share of first preference votes (24.5 percent). The outgoing governing party Fine Gael was relegated to third place (20.9 percent), while Fianna Fáil's vote share declined by 2.1 percent to 22.2 percent. In terms of Dáil seats, the outcome was close to a three-way tie. Fianna Fáil ended up marginally the largest party with 38 seats, just one seat ahead of Sinn Féin on 37 seats. Fine Gael ended up with 35 seats. The Green Party quadrupled its seat share to 12.

After months of negotiations, a new government was formed in late June 2020 composed of Fianna Fáil, Fine Gael and the Green Party. The Programme for Government agreed between the three parties, contained significant commitments on climate action, including a pledge to reduce GHG emissions by 51% over the course of the decade 2021–2030, a commitment to introduce a Climate Action (Amendment) Bill within the first 100 days of the government, and significant pledgese on sustainable transport, renewable energy and retrofitting (Government of Ireland 2020). However, the new government's agenda was criticised by many on the political left particularly for its approach to high-salience issues such as housing, homelessness and health. Indeed, data from opinion polls conducted during the campaign as well as from the exit polls showed clearly that general election 2020 did not turn out to be the climate election that some expected. The—for many, more immediate—concerns of housing and health dominated the election campaign. Whether the period from 2017 onwards represents a true transformation of Ireland's approach to climate change or merely a short-term deviation from a longer term trajectory remains to be seen.

## Bibliography


Assembly, Citizens'. 2018. *Third Report and Recommendations of the Citizens' Assembly: How the State Can Make Ireland a Leader in Tackling Climate Change*. Citizens' Assembly: Dublin.

Barry, J. 2009. '"It Ain't Easy Being Green": Sustainable Development Between Environment and Economy in Northern Ireland. *Irish Political Studies* 24 (1): 45–66.

Barry, F., A. Hannan, and E.A. Strobl. 1999. The Real Convergence of the Irish Economy and the Sectoral Distribution of Employment Growth. In *Understanding Ireland's Economic Growth*, ed. F. Barry. Basingstoke: Macmillan.

Brennan, P., and J. Curtin. 2008. *The Climate Change Challenge: Strategic Issues, Options and Implications for Ireland*. Dublin: Institute of International and European Affairs.

Burck, J., F. Marten, C. Bals, and N. Höhne. 2017. *The Climate Change Performance Index: Results 2018*. Berlin: Germanwatch.

Burck, J., et al. 2018. *The Climate Change Performance Index: Results 2019*. Berlin: Germanwatch.

Carter, N., C. Little, and D. Torney. 2019. Climate Politics in Small European States. *Environmental Politics* 28 (6): 981–996.

CCAC. 2018. *Annual Review 2018*. Dublin: Climate Change Advisory Council.

———. 2019. *Annual Review 2019*. Dublin: Climate Change Advisory Council.

Coghlan, O. 2007. Irish Climate-Change Policy from Kyoto to the Carbon Tax: A Two-Level Game Analysis of the Interplay of Knowledge and Power. *Irish Studies in International Affairs* 18: 131–153.

Convery, F.J., L. Dunne, and D. Joyce. 2013. Ireland's Carbon Tax and the Fiscal Crisis: Issues in Fiscal Adjustment, Environmental Effectiveness, Competitiveness, Leakage and Equity Implications. *OECD Environment Working Papers 59*. Paris: Organisation for Economic Co-operation and Development.

CSO. 2020. National Income and Expenditure Annual Results. Available at: https://www.cso.ie/en/statistics/nationalaccounts/nationalincomeandexpenditureannualresults/. Accessed 25 Feb 2020.

Cunningham, P. 2008. *Ireland's Burning*. Dublin: Poolbeg Press.

Curtin, J., and T. Arnold. 2016. *A Climate-Smart Pathway for Irish Agricultural Development. Exploring the Leadership Opportunity*. Dublin: Institute of International and European Affairs.

Curtis, J., and S. Lyons. 2011. Managing Household Waste in Ireland: Behavioural Parameters and Policy Options. *Journal of Environmental Planning and Management* 54 (2): 245–266.

DCCAE. 2017. *National Mitigation Plan*. Dublin: Department of Communications, Climate Action and Environment.

———. 2018a. *Local Authority Adaptation Strategy Development Guidelines*. Dublin: Department of Communications, Climate Action and Environment.

———. 2018b. *National Adaptation Framework*. Dublin: Department of Communications, Climate Action and Environment.

———. 2018c. *Sectoral Planning Guidelines for Climate Change Adaptation*. Dublin: Department of Communications, Climate Action and Environment.

———. 2019. *Climate Action Plan to Tackle Climate Breakdown*. Dublin: Department of Communications, Climate Action and Environment.

———. 2020. General Scheme Climate Action Amendment Bill. Available at: https://dccae.gov.ie/en-ie/climate-action/legislation/Pages/General-Scheme-Climate-Action-Amendment-Bill.aspx. Accessed 25 Feb 2020.

DCENR. 2015. *Ireland's Transition to a Low Carbon Energy Future, 2015–2030.* Dublin: Department of Communications, Energy & Natural Resources.

DCMNR. 2007. *Delivering a Sustainable Energy Future for Ireland.* Dublin: Department of Communications, Marine and Natural Resources.

DECLG. 2012. *National Climate Change Adaptation Framework.* Dublin: Department of Environment, Community and Local Government.

DEHLG. 2007. *National Climate Change Strategy 2007–2012.* Dublin: Department of Environment, Heritage and Local Government.

DELG. 2000. *National Climate Change Strategy Ireland.* Dublin: Department of Environment and Local Government.

Devaney, L., and D. Torney. 2019. *Advancing the Low Carbon Transition in Irish Transport.* Dublin: National Economic and Social Council.

DTTAS. 2009. *Smarter Travel—A Sustainable Transport Future.* Dublin: Department of Transport, Tourism and Sport.

Ecologic. 2020. *Climate Laws in Europe: Good Practices in Net-Zero Management.* Berlin: Ecologic Institute.

EPA. 2013. *Ireland's Greenhouse Gas Emissions in 2012.* Wexford: Environmental Protection Agency.

———. 2019a. *Ireland's Greenhouse Gas Emissions Projections, 2018–2040.* Wexford: Environmental Protection Agency.

———. 2019b. *Ireland's Provisional Greenhouse Gas Emissions, 1990–2018.* Wexford: Environmental Protection Agency.

European Commission. 2019. *Communication from the Commission: The European Green Deal.* Brussels: European Commission.

Eurostat. 2020. Population Density. Available at: https://ec.europa.eu/eurostat/databrowser/view/tps00003/default/table?lang=en. Accessed 25 Feb 2020.

Flynn, B. 2003. Much Talk but Little Action? 'New' Environmental Policy Instruments in Ireland. In *New' Instruments of Environmental Governance? National Experiences and Prospect*, ed. A. Jordan, R.K. Wurzel, and A.R. Zito, 137–156. London: Frank Cass.

———. 2006. *The Blame Game: Rethinking Ireland's Sustainable Development and Environmental Performance.* Dublin: Irish Academic Press.

Gilardi, F. 2005. The Formal Independence of Regulators: A Comparison of 17 Countries and 7 Sectors. *Swiss Political Science Review* 11 (4): 139–167.

Government of Ireland. 2010. *National Renewable Energy Action Plan—Ireland.* Dublin: Government of Ireland.

———. 2014. *Climate Action and Low-Carbon Development: National Policy Position Ireland.* Dublin: Government of Ireland.

———. 2020. *Programme for Government: Our Shared Future.* https://www.finegael.ie/app/uploads/2020/06/ProgrammeforGovernment_Final_16.06.20-1.pdf.

Harrison, K. 2010. The Comparative Politics of Carbon Taxation. *Annual Review of Law and Social Science* 6 (1): 507–529.

Hickey, K. 2008. *Five Minutes to Midnight? Ireland and Climate Change*. Belfast: White Row Press.

———. 2010. *Deluge: Ireland's Weather Disasters 2009–2010*. Dublin: Open Air.

Houses of the Oireachtas. 2019. *Report of the Joint Committee on Climate Action. Climate Change: A Cross-Party Consensus for Action*. Dublin: Houses of the Oireachtas.

IPCC. 2018. *Global Warming of 1.5 °C*. Geneva: Intergovernmental Panel on Climate Change.

Kirby, P. 2013. Policy Optimism: NESC, Climate Change and Achieving Decarbonisation. *Administration* 61 (2): 75–90.

Ladrech, R., and C. Little. 2019. Drivers of Political Parties' Climate Policy Preferences: Lessons from Denmark and Ireland. *Environmental Politics* 28 (6): 1017–1038.

Laffan, B., and J. O'Mahony. 2008. Bringing Politics Back In'. Domestic Conflict and the Negotiated Implementation of EU – Nature Conservation Legislation in Ireland. *Journal of Environmental Policy & Planning* 10 (2): 175–197.

Little, C. 2017. Intra-Party Policy Entrepreneurship and Party Goals: The Case of Political Parties' Climate Policy Preferences in Ireland. *Irish Political Studies* 32 (2): 199–223.

Little, C., and D. Torney. 2017. The Politics of Climate Change in Ireland: Symposium Introduction. *Irish Political Studies* 32 (2): 191–198.

Lockwood, M., C. Kuzemko, C. Mitchell, and R. Hoggett. 2017. Historical Institutionalism and the Politics of Sustainable Energy Transitions: A Research Agenda. *Environment and Planning C: Politics and Space* 35 (2): 312–333.

MacCarthaigh, M. 2011. Politics, Policy Preferences and the Evolution of Irish Bureaucracy: A Framework for Analysis. *Irish Political Studies* 27 (1): 23–47.

McGloughlin, J.S., and J. Sweeney. 2011. Multi-Level Climate Policies in Ireland. *Irish Geography* 44 (1): 137–150.

NESC. 2012. *Ireland and the Climate Change Challenge: Connecting 'How Much' with 'How To'*. Dublin: National Economic and Social Council.

———. 2014. *Wind Energy in Ireland: Building Community Engagement and Social Support*. Dublin: National Economic and Social Council.

———. 2019. *Climate-Change Policy: Getting the Process Right*. Dublin: National Economic and Social Council.

Ó'Broin, D., and P. Kirby. 2015. *Adapting to Climate Change: Governance Challenges*. Dublin: Glasnevin Publishing.

O'Mahony, T. 2013. Decomposition of Ireland's Carbon Emissions from 1990 to 2010: An Extended Kaya Identity. *Energy Policy* 59: 573–581.

O'Mahony, P., and K. Keohane. 2011. *Irish Environmental Politics After the Communicative Turn*. Manchester: Manchester University Press.

O'Sullivan, K.P.V., and T. Kennedy. 2010. What Caused the Irish Banking Crisis? *Journal of Financial Regulation and Compliance* 18 (3): 224–242.

OECD. 2020. OECD Data: Value Added by Activity – Ireland. Available at: https://data.oecd.org/natincome/value-added-by-activity.htm. Accessed 25 Feb 2020.

Taylor, G. 1998. Conserving the Emerald Tiger: The Politics of Environmental Regulation in Ireland. *Environmental Politics* 7 (4): 53–74.

Taylor, G., and A. Horan. 2001. From Cats, Dogs, Parks and Playgrounds to IPC Licensing: Policy Learning and the Evolution of Environmental Policy in Ireland. *The British Journal of Politics & International Relations* 3 (3): 369–392.

Teagasc. 2020. Agriculture in Ireland. Available at: https://www.teagasc.ie/rural-economy/rural-economy/agri-food-business/agriculture-in-ireland/. Accessed 26 Feb 2020.

Torney, D. 2017. If at First You Don't Succeed: The Development of Climate Change Legislation in Ireland. *Irish Political Studies* 32 (2): 247–267.

———. 2018. *Enabling Decarbonisation: A Study of Energy Sector Governance in Ireland*. Wexford: Environmental Protection Agency.

———. 2019. Climate Laws in Small European States: Symbolic Legislation and Limits of Diffusion in Ireland and Finland. *Environmental Politics* 28 (6): 1124–1144.

Torney, D., and R. O'Gorman. 2019. A Laggard in Good Times and Bad? The Limited Impact of EU Membership on Ireland's Climate Change and Environmental Policy. *Irish Political Studies* 34 (4): 575–594.

SECTION II

# Law and Policy

# Climate Litigation, Politics and Policy Change: Lessons from Urgenda and Climate Case Ireland

*Sadhbh O Neill and Edwin Alblas*

## Introduction

A Deliveroo bag at the courtroom entrance, parents with babies, school children doing their homework and a range of interested members of the public, all sitting on the floor due to a lack of space. These are not the usual sights in your average court hearing, but were telling of the way in which the Irish environmental non-governmental organisation (ENGO) Friends of the Irish Environment was able to draw broad public appeal for its 'Climate Case' against the Irish government.[1] The atmosphere during the hearing of the case in the Four Courts in Dublin from 22 to 25 January 2019 was described in the media as 'festival-like' (Sargent 2019). The case—the first of its kind in the country—however dealt with an issue of utmost gravity. In short, it centred on the Irish State's alleged failure to

S. O Neill (✉) • E. Alblas
School of Politics and International Relations, University College Dublin, Dublin, Ireland
e-mail: sadhbh.o-neill@ucdconnect.ie; edwin.alblas@ucdconnect.ie



D. Robbins et al. (eds.), *Ireland and the Climate Crisis*, Palgrave Studies in Media and Environmental Communication,
https://doi.org/10.1007/978-3-030-47587-1_4

comply with national, international and human rights law by adopting a National Mitigation Plan (NMP) that failed to set Ireland on a pathway aligned with the latest climate science as articulated by the Intergovernmental Panel on Climate Change (IPCC) and international and European Union (EU) law. On 19 September 2019, the High Court concluded that, while Friends of the Irish Environment had a right to take the case, the Government's adoption of the contested Mitigation Plan was not unlawful.

Climate Case Ireland is now only one of the two global cases in which the highest national court of law has required a Government to increase the ambition of the national climate policy, together with the Dutch 'Urgenda' case that is often described as a model and inspiration of a successful climate litigation action (Peel and Osofsky 2018).[2] As will be described in more detail in section 'Urgenda: Case Fundamentals and Policy Impacts', the ENGO Urgenda was successful in its 2015 case and subsequent appeals by the State, resulting in an order for the Dutch government to reduce its greenhouse gas emissions by 25% before the end of 2020, as compared to 1990 (Verschuuren 2019; Loth 2018). Focusing on the policy dimensions of climate litigation, this chapter will critique Climate Case Ireland and the current climate litigation archetype Urgenda. Doing so, it will be explained that while the two cases have had different outcomes to date, in both cases climate litigation has proven to be an important tool to push climate change into a central position in the legal sphere and also in the political domain. To analyse the full impact of a climate litigation action, we argue, therefore, that it is important to look not only at the legal outcome but also at the way in which the litigation effort as such shapes political discourse and the policy agenda. Basing our findings on an extensive analysis of the two respective cases, this chapter provides readers with a closer understanding of interaction between climate litigation, politics and policy change.

## Climate Litigation and Policy Change: An Introduction

In the last three decades, both Ireland and the Netherlands—in line with the rest of the EU Member States—committed to a range of international climate agreements, including the Kyoto Protocol (1997) and the Paris Agreement (2015). Such obligations exist alongside a range of other, more general, human rights obligations that countries are required to uphold and which are increasingly coming to the forefront in climate politics. At the same time, the nascent global climate regime has been subject to a catalogue of policy failures, reversals and poor implementation by states globally, as a result of which only minimal action has been taken to date to tackle the growing severity of climate risks (IPCC 2018). In this context, concerned citizens and civil society organisations are increasingly shifting their focus from the political domain to the courts, seeking legal clarifications, remedies and enforcement of countries' legal obligations in relation to climate change (Burgers 2020; Peel and Osofsky 2018).

Globally, climate litigation has 'exploded' in recent years (Peel and Osofsky 2018), with over 1000 cases currently listed on the Climate Change Litigation Database, taken against a range of actors (e.g. states or private companies), on the basis of an array of legal grounds (e.g. administrative, civil, human rights based), and with diverging goals (e.g. reparation, mandating climate action) (Burgers 2020). There are potentially many different motivations behind climate litigation: the claimant may, for instance, seek to assign responsibility for climate change to corporations or states, or seek redress for a 'wrong' (e.g. loss and damage) (O'Neill 2019). Alternatively, litigants may seek the clarification of duties of public bodies under existing climate law (e.g. Urgenda and Climate Case Ireland). Other actions seek to first and foremost put climate change on the political agenda (e.g. Inuit petition to Inter American Commission on Human Rights in 2005) or to expand legal opportunity structures to take cases in the first place (e.g. People's Climate Case in the Court of Justice of the EU). Similarly, a motivation for litigation can be the mobilisation of a community likely to be impacted or of the citizenry as a whole in an effort to publicise—and politicise—climate policy choices.

As a general observation, many of the legal arguments deployed by litigants appeal to the scientific authority of the IPCC to connect issues such as climate impacts or risks and time frames across state boundaries (e.g. the German case Lyuia v RWE AG) or to assign clear causal responsibility for climate harms or for inadequate policy responses. The United States, in particular, has seen a number of historically important actions being instigated against the fossil fuel industry, based firmly in climate change attribution science and following the model of the successful class actions against the tobacco industry and asbestos manufacturers, though the outcomes of these cases are still some years away.

It is important to note that many of the cases to date encounter familiar legal obstacles associated with procedural and fundamental rights, particularly relating to access to justice, as well as a reluctance of the courts to interfere with the policy functions of the executive in light of the 'separation of powers'. Furthermore, in considering standards or thresholds for the assessment of risk, the courts may be increasingly expected to engage with complex climate science—a task they might not necessarily be equipped for as of yet and may be reluctant to exercise (Otto et al. 2017).

Summing up, if conventional politics does not work, and political parties do not respond to the rising salience of climate change risks, mobilised citizens may direct their efforts through litigation. Burgers (2020) compares climate litigation herein with acts of civil disobedience (e.g. *à la* Extinction Rebellion). While the latter serves to invoke the law by breaching it, both climate litigation and civil disobedience point to potential injustices in how rules are made and interpreted. In this context, the legal system becomes a vehicle for advocates of urgent climate action to bypass obstacles within the policy process and to demand effective responses. It is an open question, however, whether the courts are the appropriate forum for issues relating to climate policy. Court mandated emission cuts are not likely to have the same democratic legitimacy as measures introduced by the (democratically elected) legislative branch. Furthermore, governments may be reluctant to follow court orders in a cooperative fashion, arguing instead that the judiciary should not interfere with distributive questions that are in the remit of the legislator. Below, we take a closer look at both the Urgenda case and Climate Case Ireland, as two examples of climate litigation efforts, taken with the aim of influencing climate policy through legal challenges.

## Urgenda: Case Fundamentals and Policy Impacts

### *Urgenda Before the Courts*

Urgenda, standing for 'urgent agenda', is a Dutch Environmental NGO that strives to contribute to the swift transition to a sustainable society. In 2013, the ENGO launched its case against the Dutch State, arguing that the Government was exercising 'hazardously negligence' by not cutting greenhouse gases in line with climate science. The ENGO sought a court order for a cut between 25% and 40% by the end of 2020, compared to 1990 emission levels. In 2015, in an unprecedented outcome, Urgenda, and the 886 Dutch citizens that acted as co-plaintiffs, won the case. Central to the proceedings was the reduction target for developed nations of 25–40% by 2020 over 1990 levels, stipulated in the 2007 Fourth Assessment Report of the IPCC. Whilst in 2007 the Netherlands had adopted a reduction target of 30%, in 2011 the Government indicated this target would not be met and that it would aim for 14–17% reduction instead. Through its judgement, the Court ordered ramped-up ambitions of at least 25% reductions by the end of 2020, as compared with 1990.

The State appealed the District Court's decision, arguing that the judge's decision to set a specific reduction target conflicted with the principle of separation of powers, through which the design of climate policies falls within the prerogative of the government of the day. Urgenda also filed an appeal, arguing it should be able to rely on human rights provisions stipulated in the European Convention on Human Rights (ECHR)—an argument that had been rejected by the District Court. The reduction target set in the 2015 District Court decision was binding from the outset, thus not subject to confirmation in appeal. Furthermore, the 25% target constituted an obligation to achieve the set result, meaning it is not enough to simply put in a certain amount of 'effort'.

In 2018, The Hague Court of Appeal confirmed the District Court's decision, relying this time, however, directly on Articles 2 (right to life) and 8 (right to family life) of the European Convention of Human Rights as the legal bases for its conclusion:

> [T]he Court believes that it is appropriate to speak of a real threat of dangerous climate change, resulting in the serious risk that the current generation of citizens will be confronted with loss of life and/or a disruption of family life (par. 45 judgment).

This shift of focus in its judgement from Dutch civil code provisions to international human rights law established Urgenda as a model for other climate litigation action across the globe. Closer to home, however, the District Court's decision was met by the decision of the Dutch state to appeal the decision before the Supreme Court. Doing so, the Government acknowledged that the risks posed by climate change are real but that it is the prerogative of the State to decide the pace and time frame in which action is taken, emphasising that the judgement reduced its policymaking discretion to too great an extent.

On 20 December 2019, the judgement of the Court of Appeal was upheld by the Supreme Court. Taking stock of the main judicial contributions to national and international climate litigation of this final judgement, three points are especially noteworthy. The first of these relates to standing to bring a case before the judge. The judge agreed that Urgenda, as an ENGO representing collective interests, can establish standing under the Dutch civil code. In environmental proceedings in particular, the judge argued, the pooling of collective interests—here specifically those interests relating to the taking of appropriate measures against the threat of dangerous climate change—serves to 'promote efficient and effective legal protection' for the benefit of citizens (par. 5.9.1). Beyond Dutch procedural law, however, the judge stressed that access to justice to defend such collective interests in the environmental realm is also mandated by the Aarhus Convention on access to information, public participation in decision-making and access to justice in environmental matters, to which not only the Netherlands but also all EU Member States as well as the EU in its own right are bound.

Secondly, in making its claim, the judge confirmed that Urgenda could rely directly on Article 2 (right to life) and Article 8 (right to family life) of the ECHR. The judge reiterated that Articles 2 and 8 of the ECHR contain positive obligations, which the State is obliged to consider in relation to climate action. Importantly, it then confirmed the connection of those positive obligations and the concrete reduction target of 25%, which, according to the judge, had come to exist as an internationally endorsed 'common ground' for the absolute minimum emission reduction required for the developed countries such as the Netherlands (par. 6.3).

Finally, the issue of responsibility is important to mention. On this issue, the Dutch government argued that the Netherlands, as a relatively small contributor to the global climate problem, should not be held independently responsible. The court found that the fact that other countries fail to meet their obligations does not release the Dutch state from its individual responsibility and that 'each reduction of greenhouse gas emissions has a positive effect on combating dangerous climate change [and] no reduction is negligible'. This is an important iteration, as similar claims against individual responsibility are also being relied upon in, for instance, the abovementioned *Lliuya v RWE* case, as well as a recent case taken by Greenpeace against the Norwegian government (Nollkaemper and Burgers 2020). As such, the recognition of each State's responsibility for tackling this collective problem may set the tone for future climate cases.

### *The Dutch Government's Policy Response*

The Dutch government's policy response to the subsequent Urgenda rulings can be described as one of appealing and delaying. After the first

Urgenda judgement of 2015, the conservative-liberal government in power rejected a legislative proposal by a Green opposition party aimed at implementing policies to meet the 25% targets. Instead, it was decided to await the appeal decision to obtain further legal clarity, as well as further scientific analysis of the current emission trajectory by the Dutch Environmental Assessment Agency.

The 2016 figures subsequently published by the Environmental Assessment Agency presented a 'win' for the Government: whereas previously a realised reduction of only 17% had been projected, the Netherlands now appeared to be on track to a reduction of 23% by 2020. This figure had a substantial margin of error, however, so that the actual reductions could turn out to be far lower than expected. From that point onwards, the Government used '4 megatons' (the additional 2% to get to 25%) as the factual reduction target. Three policy plans were developed to meet this target, namely, saving energy combined with producing more renewables, storing $CO_2$ emissions from one of the main coal plants underground and reducing phosphate output from dairy production. Whether these measures would, in fact, result in a reduction of four megatons was never calculated, nor were these measures implemented in practice. In 2017 and 2018, however, it became increasingly clear that substantial $CO_2$ reductions were not being realised, mainly because the economy grew more rapidly than projected, and renewable energy initiatives did not materialise. While it awaited the second Urgenda judgement—likely anticipating a quash by the appeal judge—the Government decided not to implement substantial climate action policies.

When, in its decision of 2018, the Appeal Court upheld the 2015 Urgenda judgement that mandated emission reductions of 25% by the end of 2020, the Government again decided to wait for new calculations from the Environmental Assessment Agency. In January 2019, a new report showed that at the state's current reduction pace, at best a reduction of 21% would be realised. To comply with the 2015 judgement, an additional nine megatons of reductions would be needed.

In March 2019, the Government closed a coal plant in Amsterdam, which had been in operation since 1953. This closure, estimated to save 1.7 megatons of $CO_2$ emissions, was initially planned for 2024 and was explicitly carried out to meet the terms of the Urgenda judgement. Prime Minister Rutte, asked by the media about the state's compliance with the Urgenda judgement, stressed that 'the goal is to meet the goal'. While the deadline was coming into view, however, further options available to significantly reduce emissions in a short time span became scarcer and seemed to enjoy little popularity with voters. The Government, in turn, declared that it would only take measures that would have wide societal support, would be cost-efficient and would fit within long-term climate plans.

With important provincial elections happening around the same period of March 2019, the political will to implement immediate climate measures simply did not materialise. Instead, a governmental task force was set up to develop a 'Climate Agreement' that would set the course for emission reductions leading up to 2030. The process of political compromise in forming this Agreement was complex, with each of the coalition parties having a set of non-negotiable policy options, vetoing, for instance, any measure relating to infrastructure or the agricultural sector. The risk of igniting public opposition further pushed the Government into an effective standstill, especially after the botched attempt by the French government to introduce a carbon tax that—combined with tax cuts for the richest 1%, amongst others—led to a wave of 'Yellow Vest' wave of protests.

When the Climate Agreement was presented on 28 June 2019, vice-Prime Minister De Jonge acknowledged that the measures stipulated therein would not be sufficient to meet the Urgenda target, emphasising instead that the Appeal Court had 'asked for the impossible' by requiring even more rigorous and ambitious mitigation than what was already being done, in such a short period of time. It is worth emphasising that the first judgement was delivered in 2015 (thus giving the Government five years to meet the reduction target) and that the State installed three new coal plants in the period after that judgement.

Overall, the Court's order to cut emissions by 25% by the end of 2020 was met with reluctance and resistance by the Government, with few policy changes implemented as of yet. If the Netherlands fails to meet the Urgenda target, the ENGO may decide to go back to court to ask for an order to enforce the judgement, for instance, through the setting of a fine for non-compliance. Looking further ahead, however, the Urgenda judgement has the potential to impact policymaking far beyond the national 25% reduction target that was set, most particularly through the way in which the judge provided an authoritative recognition of the close connection between climate change and concrete human rights obligations. This, in turn, may be used as a basis for future climate cases as well as policy discourse across the globe.

## Climate Case Ireland: Case Fundamentals and Policy Implications

### *Climate Case Ireland Before the Court*

Inspired by the Urgenda case, the ENGO Friends of the Irish Environment (FIE) launched a Judicial Review of the National Mitigation Plan in the High Court in 2017. The plan was chosen as the vehicle for the legal action on the basis that it was required by law to contribute to the National Transition Objective of reducing emissions from the non-land-use sectors by 80% by 2050. FIE argued that the plan, as adopted, failed to achieve this requirement and pointed out that the Irish State is fully on course to miss EU 2020 and 2030 climate and energy targets.

Despite a series of plans dating from the 2000s and new legislation introduced in 2015, Irish emissions have not yet fallen below their 1990 levels contrary to Ireland's obligations under 1992 UN Framework Convention on Climate Change (see chapter "Ireland's Policy Response to Climate Change: An Historical Overview" by Torney and chapter "The Party Politics of Climate Change in Ireland" by Little, this volume). After a decline during the recession, emissions are now projected to continue to increase until the end of this decade (EPA 2019). The 2015 Climate Action and Low Carbon Development Act, when finally enacted after years of debate and numerous drafting proposals, was weak and unambitious, leading some observers to suggest that it might be better understood as a piece of 'symbolic legislation', effectively 'designed not to work' (Torney 2019; see also chapter "Climate Law in Ireland: EU and National Dimensions" by O'Gorman, this volume). The National Mitigation Plan (NMP) of 2017 was supposed to set out a pathway for a low-carbon transition by 2050, but did not depart radically from existing policy, nor was it backed up with tangible policy measures or instruments to effect emission reductions from agriculture, buildings and transport, all sectors from which emissions have been increasing since the beginning of the economic recovery in 2013. Similarly, the Government's flagship plan *Project Ireland 2040* did not attempt to quantify the emissions implications of infrastructural investments.

When the case was heard in the Four Courts from 22 to 25 January 2019, FIE argued that the NMP breached the Climate Action and Low Carbon Development Act 2015 and, in failing to set Ireland on a course of necessary decarbonisation in order to prevent the worst effects of climate change, violated human rights. Unlike the Dutch case, the Irish climate case was argued under administrative law provisions, with FIE asking the judge to conduct a judicial review of the legality of the government's adoption of NMP. FIE argued that the Plan as adopted was unlawful, should be rewritten in line with the law and re-presented to the Oireachtas. FIE did not ask the Court to specify any emission reduction target or specific climate policies.

The High Court delivered its final judgement on 19 September 2019, finding against FIE. However, the High Court did find that FIE, established as a company limited by guarantee, was entitled to argue that the Plan was in breach of the 2015 Act and even to plead that the Plan breached human rights. The judgement also referred to another case where FIE had challenged the decision by Fingal County Council to grant an extension of planning permission for a new runway at Dublin airport, arguing inter alia, that the runway would give rise to additional emissions, which would impact on the unenumerated constitutional right of Irish citizens to an environment. While the challenge ultimately failed, the Court did agree that there was indeed:

> [a] right to an environment that is consistent with the human dignity and well-being of citizens at large… [Such a] right "is an essential condition for the fulfilment of all human rights. It is an indispensable existential right that is enjoyed universally, yet which is vested personally as a right that presents and can be seen always to have presented, and to enjoy protection, under Article 40.3.1° of the Constitution". [Furthermore, it] is not so utopian … that it can never be enforced…. (FIE v Fingal County Council [2017] IEHC 695 par. 245)

Despite some academic debate over the unenumerated rights doctrine generally (Kelleher 2018), an interpretation of the relevance and force of the Barret, J. judgement in the runway case was not developed in the climate case judgement.

FIE was successful in seeking leave to appeal before the Supreme Court, which was heard on 22 and 23 June 2020. In a ground-breaking judgment delivered on the 31st of July 2020, seven Supreme Court judges found unamiously that the 2017 fell 'well short' of the specificity that would be required to comply with the 2015 Act. The Court quashed the existing Plan, requiring the government to formulate a new plan that covers the period remaining to 2050. While the detials of this plan are not known at time of publication, it is clear that the decision in the case has major implications for climate law in terms of clarifying the precise duties of the Minister under the 2015 Act, the degree of discretion available to the executive in formulating climate policy, and the justiciability of national mitigation plans. At the same time, the rights argument advanced in the case under the Irish Constitution and the European Convention on Human Rights was less successful. With regard to the unenumerated right to a healthy environment, the Chief Justice concluded that such a right cannot be derived from the Constitution, hinting however that existing constitutional rights and state obligations may be nonetheless relevant in environmental litigation. In light of the above, the case may be described as politically 'disruptive' (Fisher et al. 2017).

### *The Irish Government's Policy Response*

The Irish Climate Case came at a time of unprecedented media coverage of and focus by the legislature on climate change, after a sharp drop in media coverage following the Paris Agreement (see chapter "Climate Change and the Media" by Robbins, this volume). The publication of the IPCC special report on the anticipated effects of global temperature rises of 1.5° (IPCC 2018) drew attention to the short window of opportunity in which emissions reductions could avert dangerous climate change and recommended emission cuts in the order of 45% by 2030. Unlike previous IPCC publications, this report and its stark warnings about the dangers of global warming had the effect of focusing minds and energising a new public that began to demand accelerated climate action, including a new youth-led international climate movement of school strikers.

In addition to the growing salience of climate change, policy innovation in the national parliament (the Dáil and Seanad, together known as the Houses of the Oireachtas) had also overtaken the slow and reluctant government response. In March 2019, after months of deliberation and negotiations, the Joint Oireachtas Committee on Climate Action published its cross-party consensus report,[3] which was heralded as 'remarkable' (Curtin 2019) for its depth and detailed recommendations for action, including a recommendation to revise the 2015 Climate Act and radically strengthen the governance and accountability mechanisms (see chapter "Climate Law in Ireland: EU and National Dimensions" by O'Gorman, this volume). This committee had been established to consider the report of the Citizens' Assembly, which had met over two weekends in 2017 to consider climate change, but whose 13 headline recommendations had not yet been adopted by the Government. Events such as strikes, litigation, reports and more grim data relating to Irish emission trajectories, all put pressure on the Government to act.

Upon his appointment in October 2018, the new Minister for Communications, Climate Action and the Environment, Richard Bruton TD, announced that he would prepare a new All of Government Climate Action Plan, which was finally published in June 2019. While the Government had argued in Court that the 2017 Plan under challenge was a 'living document', it is significant that in the space of six months following his appointment, an entirely new plan was drawn up and approved by the Government. It is reasonable to presume that the prospect of losing the High Court case against a growing tide of criticism over the State's climate policies generally helped to build the case for updating Irish climate policy. In contrast to the NMP, whose targeted abatement measures were referred to by FIE as 'weak and unambitious' and not calculated to achieve the emissions reductions that would be required in order for Ireland to achieve a transition to a low-carbon economy by 2050, the new plan had 183 concrete actions built around a specially commissioned economic analysis.

The publication of the 2019 Plan did mark a watershed in climate policy, even though it relies heavily on the take-up of electric vehicles and market-driven solutions instead of a redirection of capital investment and deeper policy reforms. It was the first comprehensive climate policy statement that quantified the emission reduction potential of various policy measures and the first time the agriculture sector had been allocated an 'indicative' emissions cap for that sector by 2030.

## Discussion

As was explained in section 'Climate Litigation and Policy Change: An Introduction' of this chapter, in the last three decades, both Ireland and the Netherlands have committed to a range of emission reduction targets. These have been negotiated alongside existing human rights obligations, where the link to climate change matters is increasingly being recognised in both the courts as well as academic scholarship (Peel and Osofsky 2018; Burgers and Staal 2018). The trouble with greenhouse gases and their impact on the atmosphere is that only substantial and immediate reductions in volumes of gases emitted will make a difference to our chances of avoiding dangerous climate change—mere commitments, even legally binding ones, are not enough. In this context, climate litigation may play a key role in enforcing a country's obligations and may succeed in holding governments and public bodies to account to ensure cross-sectoral policy integration and compliance with emission budgets or targets.

In Ireland, in particular, policy failure is in no small way due to the poor legislative framework that underpins climate policy in Ireland. Critics have highlighted the importance of revising the 2015 Act (see chapter "Climate Law in Ireland: EU and National Dimensions" by O'Gorman, this volume), which also happens to be a commitment in the Government's Climate Action Plan of 2019, and in the Programme for Government adopted by the new coalition government in June 2020. Such a revision will need to enhance the current legal framework for effective climate governance by introducing more ambitious targets consistent with the Paris Agreement goals, and new and strengthened forms of accountability for climate policy. Climate *politics*, on the other hand, will likely reflect broader anxieties about any large-scale curtailment of economic activities such as livestock farming, along with potential increases in the cost of living due to carbon taxes.

There are precedents for paradigmatic changes, however: Ireland has recent experience of litigation-driven constitutional and policy reforms, when an unsuccessful legal action relating to the tax status of same-sex couples eventually led to the Marriage Equality referendum in 2015.[4] Losing in the courts might still mean eventual victory in the policy arena, if a case builds sufficient momentum and garners enough political will to drive bold decisions and see them through. The Dutch experience, however, shows that even a solid legal victory does not guarantee emission cuts or an effective political response on the scale of what is required. In this context, it can be concluded that while climate litigation does not offer a panacea for climate inaction by states, it does provide citizens and ENGOs with a meaningful and important instrument to press for change.

## Conclusion

This chapter began by arguing that to analyse the full impact of a climate litigation action, it is important to look not only at the legal outcome, but also at the way in which the litigation effort as such shapes political discourse and the policy agenda. In the Netherlands, the order to cut emissions 25% by 2020 was met with reluctance and effectively resistance by the Dutch government, which adopted a response of 'appealing and delaying'. As a result, few policy changes have been implemented to date. In the long term, however, Urgenda has the potential to impact climate policy far beyond the national 25% reduction target that was set, most particularly, in connecting climate change and concrete human rights obligations in public discourse and in the law. As for the Irish case, the Court's quashing of the 2017 National Mitigation Plan means the Government is now legally required to increase the ambition of its national climate policy.

In sum, while litigation is a powerful instrument in the advocacy toolkit, by itself it is unlikely to be sufficient to effect the emission reductions that are recommended by the IPCC. As such, it is vital that ENGOs and civil society continue to push for climate action across the different advocacy channels discussed in this volume, including activism, community and political engagement and fostering media attention. The research agenda also needs to expand in many directions if it is to contribute usefully to the policy process. Complex socio-economic transitions require new and bold thinking, driven by evidence, innovation and imagination, along with political courage and new forms of public mobilisation and institution building. To this end, much work is still to be done across the board.

## Notes

1. Full case title: Friends of the Irish Environment vs. the Government of Ireland, Ireland and the Attorney General (19 September 2019) Ref. 2017/793JR, publicly referred to, and hereinafter denoted as, 'Climate Case Ireland'.
2. Full case title: The State of the Netherlands v Urgenda Foundation, Hoge Raad (20 December 2019) case 19/00135 (English translation); The State of the Netherlands v Urgenda Foundation, The Hague Court of Appeal (9 October 2018), case 200.178.245/01 (English translation); Urgenda Foundation v The State of the Netherlands, District Court of The Hague (24 June 2015), case C/09/456689/HA ZA 13-1396 (English translation).
3. Although it should be noted that the final report was not approved by Sinn Féin or People Before Profit due to the inclusion of a recommended increase in the carbon tax.
4. Zappone & Gilligan v. Revenue Commissioners & Ors [2006] IEHC 404.

## Bibliography

Burger, M., R. Horton, and J. Wentz. 2020. The Law and Science of Climate Change Attribution. *Columbia Journal of Environmental Law* 45 (1): 57–239.

Burgers, L. 2020. Should Judges Make Climate Change Law? *Transnational Environmental Law* 9 (1): 55–75.

Burgers, L., and T. Staal. 2018. Climate Action as Positive Human Rights Obligation: The Appeals Judgment in Urgenda v The Netherlands. In *Netherlands Yearbook of International Law, T.M.C*, ed. R.A. Wessel, W. Werner, and B. Boutin. Asser Press: The Hague, NL.

Curtin, Joe. 2019. Ireland's Radical Plan Could Stave Off Worst Climate Outcome. https://www.irishtimes.com/opinion/ireland-s-radical-plan-could-stave-off-worst-climate-outcome-1.3928503

EPA. 2019. *Ireland's Greenhouse Gas Emissions Projections 2018–2040.* Available at http://epa.ie/pubs/reports/air/airemissions/ghgprojections20182040/Greenhouse_Gas_Projections.pdf

Fisher, E., E. Scotford, and E. Barritt. 2017. The Legally Disruptive Nature of Climate Change. *The Modern Law Review* 80 (2): 173–201.

IPCC. 2018. Summary for Policymakers. In *Global warming of 1.5°C. An IPCC Special Report on the impacts of global warming of 1.5°C above pre-industrial levels and related global greenhouse gas emission pathways, in the context of strengthening the global response to the threat of climate change, sustainable development, and efforts to eradicate poverty*, eds. P. Z. Masson-Delmotte, H. O. Pörtner, D. Roberts, J. Skea, P.R. Shukla, A. Pirani, W. Moufouma-Okia, C. Péan, R. Pidcock, S. Connors, J. B. R. Matthews, Y. Chen, X. Zhou, M. I. Gomis, E. Lonnoy, T. Maycock, M. Tignor, T. Waterfield. World Meterological Organisation: Geneva, Switzerland.

Kelleher, O. 2018. The Revival of the Unenumerated Rights Doctrine: A Right to an Environment and Its Implications for Future Climate Change Litigation in Ireland. *Irish Planning and Environmental Law Journal* 25 (3): 97–103.

Loth, M.A. 2018. Too Big to Trial? Lessons from the Urgenda Case. *Uniform Law Review* 23 (2): 336–353.

Nollkaemper, A., and L. Burgers. 2020. A New Classic in Climate Change Litigation: The Dutch Supreme Court Decision in the Urgenda Case. https://www.ejiltalk.org/a-new-classic-in-climate-change-litigation-the-dutch-supreme-court-decision-in-the-urgenda-case/

O'Neill, S. 2019. 'Carbon Crimes' *Working Notes* Journal of the Jesuit Centre for Faith and Justice, 84, June. Available at: https://www.workingnotes.ie/current-issue/item/carbon-crimes

Otto, F.E.L., R.B. Skeie, J.S. Fuglestvedt, T. Berntsen, and M.R. Allen. 2017. Assigning Historic Responsibility for Extreme Weather Events. *Nature Climate Change* 7: 757.

Peel, J., and H.M. Osofsky. 2018. A Rights Turn in Climate Change Litigation? *Transnational Environmental Law* 7 (1): 37–67.

Sargent, N. 2019. 'Festival Vibe' in Court as Public Makes Stand for Climate Action. *Green News*, January 22. Available at: https://greennews.ie/festival-vibe-court-public-makes-stand-climate-action/

Torney, D. 2019. Climate Laws in Small European States: Symbolic Legislation and Limits of Diffusion in Ireland and Finland. *Environmental Politics* 28 (6): 1124–1144.

Verschuuren, J. 2019. The State of the Netherlands v Urgenda Foundation: The Hague Court of Appeal Upholds Judgment Requiring the Netherlands to Further Reduce Its Greenhouse Gas Emissions. *Review of European, Comparative and International Environmental Law* 28 (1): 94–98.

# Climate Law in Ireland: EU and National Dimensions

*Roderic O'Gorman*

*Obviously, climate emissions and greenhouse gas areas is an area where we're laggard and falling way behind*
—An Taoiseach Leo Varadkar, December 2018

## Introduction

The admission by An Taoiseach Leo Varadkar, of Ireland's failure to cut carbon emissions, vis-à-vis other European Union (EU) Member States, represents a recognition of an ongoing failure by the State to seriously tackle Ireland's contribution to the global climate crisis. Indeed, Ireland's track record with respect to environmental regulation generally has remained poor throughout our membership of the EU, irrespective of wider issues like the rapid economic development of the early 2000s or the subsequent economic crisis (Torney and O'Gorman 2019). On the face of it, the challenges Ireland faces in cutting emissions are not quantifiably greater than those faced by other European states. We are governed by the

R. O'Gorman (✉)
School of Law and Government, Dublin City University, Dublin, Ireland
e-mail: roderic.ogorman@dcu.ie



D. Robbins et al. (eds.), *Ireland and the Climate Crisis*, Palgrave Studies in Media and Environmental Communication,
https://doi.org/10.1007/978-3-030-47587-1_5

same international and EU legal regime, more recently complimented by our own set of domestic legislation on the issue of climate. And yet in 2018, Ireland was judged to have the worst record for performance on climate action within the EU, though there has been some small improvement in subsequent years (Burck et al. 2018).

This chapter seeks to map the diverse sources of climate law as applicable in Ireland and highlight why law has so far failed to secure meaningful climate action within the State. It begins by briefly charting Ireland's membership of the international climate regime. It then outlines the manner in which a bundle of binding EU legislative measures contained in the *2030 Climate and Energy Framework* impacts domestically and assesses the degree to which Ireland has benefitted from certain flexibilities built into this legislation. The chapter then analyses the provisions of Ireland's initial piece of domestic climate legislation, the *Climate Action and Low Carbon Development Act*, 2015. Having outlined the key features of this law, it analyses how the regime created compares negatively with that applicable in two EU states—the UK and Sweden—on a range of different themes. The chapter concludes by examining the heightened focus on the issue of climate change in 2019 and charts how increased public interest in the issue has enhanced its political profile in Ireland—reflected in developments such as the Citizens' Assembly, the Oireachtas Special Committee on Climate Change and the Action Plan on Climate Change—and considers whether these may portend a more rigorous domestic response to the climate crisis going forward.

## EU Climate Law as It Applies to Ireland

Ireland is a signatory to the United Nations Framework Convention on Climate Change (UNFCCC) which was signed at the Earth Summit in Rio De Janeiro in 1992. The state signed up to the Kyoto Protocol, adopted under the UNFCCC, in 1998 and ratified it in 2002. Like all other EU Member States, Ireland signed the Paris Agreement, the successor to Kyoto, and subsequently ratified in late 2016. Whereas Kyoto had only set specific emission limits for a group of developed states (including Ireland), the Paris Agreement requires all states to provide individual commitments on emissions reduction measures—Nationally Determined Contributions (NDCs)—on a five-yearly cyclical basis, with a view that the combined NDCs will reduce emissions sufficiently to limit the global temperature increase by 2050 to between 1.5 and 2 degrees.

As a Member State of the European Union, Ireland is bound by the collection of EU regulations, directives and decisions that comprises EU climate law. The policy field of 'the environment' has been a shared competence between the Union and Member States since it was first listed in the Treaties as part of the Single European Act in 1987. The competence is now enumerated through Articles 191 and 192 of the Treaty on the Functioning of the European Union (TFEU). As part of the Lisbon Treaty in 2009, new text was added that specifically named combatting climate change as one of the Union's objectives in this area. The Treaties also give the EU shared competence in the field of energy under Article 194 TFEU.

Having signed the UNFCCC and the Kyoto Protocol, the EU was seen as an 'avid campaigner for the global adoption of binding emissions targets' (Kingston et al. 2017). Its strong stance in the international climate sphere has also been matched by internal actions which have grown in ambition. In 1991, the European Commission set out its Community Strategy to Limit Carbon Dioxide Emissions and Improve Energy Efficiency. Subsequently, the First Climate Change Programme (2000) and the Second Climate Change Programme (2005) were drawn up to implement the measures needed for the EU to meet its commitments under the Kyoto Protocol. As negotiations took place to find a successor to Kyoto, which eventually evolved into the Paris Agreement, the EU's own targets in respect of climate were drawn together in two sequential packages of legislative and policy measures: the *2020 Climate & Energy Package* and the *2030 Climate & Energy Framework*.

The 2020 Package was agreed by the European Council in 2008 and was abbreviated as the 20-20-20 target: an overall reduction in EU greenhouse gas (GHG) emissions by 20% by 2020, along with achieving 20% of EU energy coming from renewables and a 20% improvement in energy efficiency by the same date. These commitments were to be enacted primarily via four legislative proposals in the energy/climate field: the Carbon Capture and Storage Directive, the Renewable Energy Directive, the Emission Trading System Directive and the Effort Sharing Decision. To meet the 20% reduction target in greenhouse gas emissions, it was decided that the sources covered by the EU's Emissions Trading System (ETS) should reduce their greenhouse gas emissions by 21% compared to 2005. The year 2005 was chosen since it was the first year for which verified emissions data for the installations covered by the ETS were available. Other emissions sources which were not covered by the ETS (e.g. transport, buildings, the services sector, small industries and agriculture) were instead included in the Effort Sharing Decision. Across these areas, a 10% reduction in their collective emissions compared to 2005 was required.

In the lead up to the Paris Agreement, and with a view to achieving the commitments that the EU proposed to enter into, headline targets were agreed for a *2030 Climate & Energy Framework* by the European Council, some of which were subsequently strengthened during negotiations between the Council and the Parliament. The 2030 Package reflects a range of more ambitious targets for the EU to achieve by 2030: a 40% cut in greenhouse gas emissions, a 32% share for renewable energy and a 32.5% improvement in energy efficiency. This would equate to a reduction of emissions from the EU ETS sectors of 43% compared to their level in 2005. For the sectors not covered by the ETS, a reduction of emissions of 30% compared to 2005 was agreed.

As well as new targets, the 2030 Framework also restructured and broadened the EU's approach. The Effort Sharing Decision was redrafted as a regulation—the Climate Action Regulation. This change to legislating via regulation created a more binding and consistent legal regime across the Member States. A key commitment of the EU contained in its Nationally Determined Contribution to the processes established under the Paris Agreement was to address emissions from the broad agriculture, land use and forestry sector. This led to the drafting of the Land Use, Land Use Change and Forestry Regulation. The Union also passed a new Energy Union and Climate Action Governance Regulation, acting to draw together a range of planning, reporting and monitoring obligations covering both climate emissions and energy issues in the context of the Energy Union.

### *Ireland and the 2030 Climate & Energy Framework*

Under the 2020 targets laid out in the Renewable Energy Directive, Ireland was obliged to achieve 16% Gross Final Consumption from renewable by 2020. Figures for 2018 show that the State remains a significant distance from this target, with only 11% of final consumption generated from renewables (Sustainable Energy Authority of Ireland 2019). While renewables represent a significant component of electricity generation, their contribution to heat and transport actually decreased. Reasons for this lack of achievement include the manner of the transposition of the EU law into domestic legislation, the (in)actions of key players in the domestic energy market and the growing conflict between the development of wind energy production capacity and environmental concerns (O'Connor 2017). Significantly, under the recast Renewable Energy Directive for 2030, mandatory national targets in the Directive are replaced by targets negotiated bilaterally between the Member State and the Commission on a more informal basis (Kingston et al. 2017).

Under the new Climate Action Regulation, Ireland is to achieve a 30% reduction in its non-EU ETS emissions by 2030, relative to 2005 emissions. Under this new legislation, Ireland benefited from two flexibilities (loopholes) provided within the Regulation. Article 6 permitted certain Member States to have EU ETS allowances collectively taken into account for their compliance under the Regulation. Ireland was listed in Annex II as such a state (one of nine) and was permitted to use up to 4% of EU ETS allowances for each year of the period from 2021 to 2030 for the purposes of calculating its compliance with the Effort Sharing Regulation.

Secondly, Ireland is the principal beneficiary of Article 7. This allows certain states to make use of the sum of total net removals and total net emissions from afforested land, deforested land, managed cropland, managed grassland and managed forest land and managed wetland, to be taken into account for its compliance under the Regulation. Ireland is entitled to use up to 26.8 million tonnes, the third highest allocation of all Member States. The Non-Governmental Organisation Transport & Environment has stated that the very existence of this flexibility is problematic and will 'just postpone measures to reduce emissions, undermining the overall ambition of the [Regulation]' (Transport and Environment 2018).

Aside from these two flexibilities, Ireland also benefits from a monitoring starting point which widely disregards the State's failure to meet its 2020 targets, something which Transport & Environment described as 'scandalous' (Transport and Environment 2018). The fact that Ireland was in the position to negotiate a particularly beneficial deal including significant concessions was seen, at least in part, as having been a consequence of the involvement of the Irish European Commissioner, Phil Hogan, who oversaw the Agriculture portfolio at this point (Irish Cooperative Organisation Society 2016).

## Domestic Climate Change Legislation

Ireland has been slow to address climate change through legislative measures (see also chapter by Torney in this volume). From the mid-2000s, piecemeal references to climate change can be found across legislation in a number of areas including emissions trading, promotion of sustainable energy, carbon tax, consumer information requirements and fuel and energy efficiency measures, though many of these laws were implementing EU regulations (Kennedy 2011). The Planning and Development (Amendment) Act 2010 included direct references to 'adaptation to climate change' and to 'anthropogenic greenhouse gas emissions'. The Act also included requirements on local authorities to carry out flood risk assessments (Linehan 2012).

The first effort to enact legislation specifically addressing climate change came about through the publication of the draft Climate Change Response Bill 2010. This law would have mandated specific emission reduction targets for 2020, 2030 and 2050, required the drawing up of a national climate change plan and the creation of an Expert Advisory Body. However, the bill fell following the collapse of the Fianna Fáil/Green Party Government in early 2011.

In 2015, Ireland finally joined the growing body of states which had adopted climate laws with the passage of the Climate Action and Low Carbon Development Act. The central element of the legislation was a 'national transition objective', whereby the State would 'pursue, and achieve, the transition to a low carbon, climate resilient and environmentally sustainable economy by the end of the year 2050'. From a policy point of view, the Act focuses on both mitigation and adaptation. It requires the drafting of a five-year National Mitigation Plan, setting out the Governments plan for achieving the national transition objective and the specific policies to be followed, taking account of obligations stemming from EU and international law. The National Mitigation Plan will be supplemented with sectoral mitigation plans, to be devised by relevant Government departments. The National Mitigation Plan must be approved by the Government and must then be laid before each House of the Oireachtas.

The Act also creates a requirement for a national adaptation framework, also drawn up on a five-yearly cycle and designed to specify the national strategy for the application of adaptation measures in different sectors and by local authorities, both to reduce the vulnerability of the State to the negative effects of climate change and to avail positive effects stemming from it. This will also be elaborated further through sectoral adaptation plans, adopted by governmental departments. Similar to the National Mitigation Plan, the national adaptation framework must also receive Government approval and be laid before the Houses of the Oireachtas.

A further element of annual accountability to the Oireachtas is created by the requirement that the Minister must present an annual transition statement. The statement, which must cover both mitigation and adaption, shall include a list of policy measures taken in that year to achieve the national transition objective, an assessment of the effectiveness of such measures, a record of greenhouse gas emissions as set out in the most recent national greenhouse gas emissions inventory prepared by the Environmental Protection Agency, a projection of future greenhouse gas emissions and a record of Ireland's compliance with EU and international legal provisions on climate.

The Act created a new, independent body—the Climate Change Advisory Council. This body is tasked with giving advice to the Government and individual ministers on the carrying out of their duties under the legislation and other actions they take to achieve climate change mitigation and adaptation. It is also obligated to conduct an annual review of all measures taken towards achieving greenhouse gas emissions reductions and furthering the transition to a low-carbon economy. This review is to be communicated to the Minister as an Annual Report. The Advisory Council can also undertake a 'periodic review' at any point if it feels there has been specific changes in the scientific knowledge or the EU or international legal regime around climate change or if it is needed to maintain progress to achieve the national transition objective. The Minister can also request the Advisory Council to undertake a periodic review.

The Council has a chairperson and eight to ten ordinary members including members representing relevant state agencies: the Environmental Protection Agency, the Sustainable Energy Authority of Ireland, Teagasc (Agriculture and Food Development Authority) and the Economic and Social Research Institute. The Act makes provision for the independence of the Council in the performance of its functions in section 11(3).

## How Does the Climate Action and Low Carbon Act, 2015 Compare Internationally?

The Climate Action and Low Carbon Act 2015 has received significant criticism as regards its scope and objectives (Torney 2017; Kennedy 2016). As domestic legislation is one area where we are not conclusively bound by EU-wide requirements, it is legitimate to question whether there is something about the national law that makes it less effective in achieving climate goals compared with that present in other Member States. In order to provide a framework to judge the effectiveness of the legislation, it is useful to compare it with legislation adopted in two other states. The British Climate Change Act 2008 is now in existence for over a decade and represents an early legislative model. The more recent Swedish Climate Policy Framework 2017 forms a central element of a national regime recognised as one of the strongest in Europe (Climate Action Network 2018; Ecologic Institute 2017). In seeking to judge the effectiveness of the Irish climate regime, five broad questions are asked: whether clear mitigation targets are set out, whether independent institutions are established, whether the system creates accountability, whether the system is insulated from political change and whether the system sends signals to the wider economy.

### *Are Clear Mitigation Targets Established?*

Under the UK Climate Change Act, a clear 2050 target is established and defined in the 'Introduction' section. This originally stated that it is the duty of the Secretary of State to ensure that the net UK carbon account for the year 2050 is at least 80% lower than the 1990 baseline, though the Act was amended in 2019 to substitute in 100% for 80%. The Act also creates a five-year budgetary cycle with interim emissions reduction targets to be achieved.

Under the Swedish legislation, a National Climate Target is established which states that by 2045, Sweden is to have no net emissions of greenhouse gases into the atmosphere and should thereafter achieve negative emissions. Net zero is defined as a reduction of GHG emissions from activities in Sweden by at least 85% compared to 1990, with the remaining reductions to zero achieved through supplementary measures. This is regarded as being one of the most ambitious targets established by national legislation. The Framework also sets out a range of targets for 2030 and 2040, both in terms of overall emissions and in certain fields, with transport specifically referenced.

Whereas the Irish legislation does create a 'National Transition Objective', defined as a low-carbon, climate resilient and environmentally sustainable economy by the end of the year 2050, this is a descriptive target, as opposed to being composed of a numerical definition. The Act does directly tie into both Article 2 of the UNFCCC and 'any mitigation commitment entered into by the EU'. However, no intermediate targets are established in the legislation. This lack of targets formed a significant part of the debate about the legislation when the Act was going through the Oireachtas (Doran 2013).

### *Are Independent Institutions Created?*

The creation of a council or committee, designed to provide advice and assessments on climate issues independent of politics, is a recurring feature of climate legislation internationally (Bauer et al. 2012). The approach of mandating an independent group to deal with controversial issues requiring difficult political decisions is not restricted to the area of climate policy. Following the economic crisis, Eurozone states were required to create such bodies where they did not exist previously as a consequence of the Fiscal Compact. This led to the formation of the Fiscal Advisory Council in Ireland, which provides assessments of fiscal and budgetary measures.

In the UK, the Committee on Climate Change was created, though it lacked the power to issue binding recommendations. This lack of rule-making capacity meant that it finds its 'influence based on reputation and authority rather than formal powers' (Lockwood 2013). Similarly, Sweden's Climate Policy Council was set up to provide independent assessments of how the overall policy presented by the Government is compatible with the national climate goals. One criticism levied at it is that its actual legal basis is contained in secondary legislation, thus, in theory at least, leaving it open to repeal without parliamentary approval (Ecologic Institute 2017).

During the debate on the Irish legislation, there was significant focus on the lack of independence of what was originally called the 'National Expert Advisory Council on Climate Change' (Doran 2013). In this author's own submission during the pre-legislative scrutiny phase, I, along with others, called for the explicit recognition of the body's independence, drawing an analogy with that clear declaration of independence contained in the Act establishing the Fiscal Advisory Council. This issue was addressed in the final legislative draft.

### *Does the Climate Change Regime Create Political Accountability?*

Accountability mechanisms are essential to ensure that governments can be challenged in the event that they are failing to reach targets. Across Ireland, the UK and Sweden, there is a common theme in how this is approached, with political accountability on an annual and medium-term basis accepted as good practice. Multi-annual plans, like the Irish Adaptation and Mitigation plans, must be laid before the national parliaments for scrutiny. Annually, a designated minister is required to outline to the national parliament the measures that have been taken within that calendar year to achieve emission reductions. In Sweden, this is done as part of the annual Budget Bill. Such an approach of an annual Carbon Budget being discussed on Budget Day had originally been adopted in Ireland under the Fianna Fáil/Green Government but was discontinued in 2011 when the subsequent Fine Gael/Labour government came to power. Under the 2015 legislation, the Minister makes an Annual Transition Statement to the Dáil. Within the UK legislation, there are specific duties placed on the relevant Secretary for State regarding the preparation of proposals and policies for meeting carbon budgets, reporting on these proposals and policies and the need for UK domestic climate action.

### *Is the Climate Change Regime Insulated from Political Change?*

One significant similarity across the three jurisdictions is the degree of political consensus regarding the legislative proposals. This cross-party backing for the means of dealing with climate change is significant in ensuring that the legislation is not subsequently undermined by successive and ideologically different governments. An example of the latter is the effort in Australia to dismantle the climate change regime that was introduced by the Labour Government in 2011. The subsequent right-wing Liberal Government succeeded in repealing the carbon tax. While it narrowly failed in its efforts to terminate the institutional structure introduced—the Climate Change Authority—this body is openly ignored by the current government and its functioning has been undermined by continuous resignations.

In the UK, the Climate Change Act 2008 had broad support across the House of Commons when it was originally passed, with only five MPs voting against the Second Reading. Prior to entering Government in 2010, the Conservative Party under David Cameron was very supportive of climate measures, but following the election win, the existing consensus broke down (Gillard 2016). Energy price increases in 2013 put significant pressure on climate measures (Lockwood 2013). Following her election as Prime Minister in 2016, Theresa May amalgamated the Department of

Energy & Climate Change, originally formed in 2009 to help deliver the carbon budgets, into the Department of Business, Energy and Industrial Strategy, thus ending distinct representation for climate issues within the Cabinet.

Sweden also saw buy-in by parties across the political spectrum, other than the far-right Swedish Democrats (Ecologic Institute 2017). It is relevant to note that the national Green Party was a part of the coalition Government in power when the legislation passed, with Green MPs holding both the Climate Change and Environment ministries.

Unlike the political situation in both the UK (Brexit Party/UKIP and significant elements of the Conservatives) and Sweden (Swedish Democrats), there is little overt climate denial in the main parties in Ireland. At the same time, there is little obvious interest in the majority of Irish political parties in the topic either (see chapter 'The Party Politics of Climate Change in Ireland' by Little, this volume). The lack of appetite to address climate issues when they are politically unpalatable was demonstrated by the Government's surprising decision not to increase the level of carbon tax as part of Budget 2019, for which it received significant press and expert criticism (O'Sullivan 2018).

### *Does the Climate Change Regime Provide Signals to the Wider Economy?*

An important consideration for any climate regime is its ability to achieve buy-in from the industrial and commercial sectors, particularly those industries which are currently of a highly carbon-intensive nature. Given that businesses need the confidence to plan for longer term investment in renewables and an understanding that measures to tackle climate change will not be once off but rather will be maintained on a sustained basis, it is vital that the structures created by climate legislation generate this certainty. If properly operated, these will allow companies the time to prepare for changes to their business model and give space to innovators.

The five-year carbon budget cycle contained in the UK legislation was designed with this in mind, particularly when combined with the requirement that carbon budgets had to be agreed 12 years before their commencement date. Lockwood (2013) notes that the greater certainty provided by the Act strengthened the hands of those in the Confederation of British Industry who saw opportunities in the area of emissions reductions. In an assessment of the Act ten years after its enactment, it was noted that the certainty introduced by the budgets had played a major role in decarbonisation in the power generation sector, but had been less successful in other parts of the economy (Fankhauser et al. 2018).

As noted above, as part of the yearly Budget Bill, the Swedish Government must submit a climate review to the Riksdag. This linkage of the announcement on carbon with the budgetary process enhances its visibility. While the Swedish law did not create a budgetary cycle as the UK act did, the fact that the legislation established a range of intermediate targets (2030, 2040, 2045) allows for some degree of business confidence.

It is difficult to see how the Irish legislation was capable of giving much of a signal to business. It contained neither a carbon budgetary cycle nor any clear intermediate targets before 2050. The fact that it tied itself so closely to EU targets means that until these are agreed, there are no targets within which Irish businesses can plan their investment decision. This is arguably the weakest element of the Irish legislation. It is submitted that it reflects inbuilt reluctance to make any moves which might see Ireland undertake higher emission reductions than the EU might subsequently prescribe.

## Implementing and Reforming the Climate Change Act

### *Actions on Foot of the New Legislation*

Following the enactment of the Climate Change and Low Carbon Development Act in 2015, the then Government initiated work on the cyclical plans required under it. The first National Mitigation Plan was published in July 2017. This was followed in January 2018 by the publication of the National Adaptation Framework. At the same time, €10 million in funding was provided for the establishment of four climate action regional offices (CARO) to facilitate greater coordination between local authorities in developing capacity for mitigation and adaptation, with a specific focus on the challenges of the specific region.

On 13 July 2016, the newly formed Government introduced a motion in the Dáil to convene a Citizens' Assembly to address a range of issues including the removal of the 8th Amendment of the Irish Constitution, how Ireland could best respond to the challenges and opportunities of an ageing population and how referendums are run. An amendment proposed by Eamon Ryan TD of the Green Party to include discussion of 'how the State can make Ireland a leader in tackling climate change' was accepted by the Dáil and the item was added to the agenda of the Citizens' Assembly. Meetings of an earlier deliberative body, the Constitutional Convention, had been seen as significant in growing public support for the 2015 referendum on permitting Marriage Equality in Ireland (Elkink et al. 2017). The Citizens' Assembly met on the topic of climate change in September and November 2017 and a final list of 13 recommendations were voted upon by it, all approved with majorities of between 80% and 100% (Citizens' Assembly 2018). Amongst these was a recommendation for new legislation to put climate change at the centre of policymaking.

In July 2018, the Dáil voted to establish a special committee of TDs and senators—the Joint Oireachtas Committee on Climate Action. Its role was to consider the report on Climate Change from the Citizens' Assembly, along with the implementation of the National Mitigation Plan, Irelands National Energy and Climate Plan (in the context of the EU Regulation on the Governance of the Energy Union) and the Sectoral Adaptation Plans. The Joint Committee published its report in March 2019 (Houses of the Oireachtas 2019) and during a subsequent Dáil debate on the report in May, an amendment whereby Ireland declared a Climate and Biodiversity emergency was passed. Ireland thus became the second country internationally to recognise a climate emergency. But while the report's title spoke of cross-party consensus, it was clear from the list of votes taken to amend sections of it that issues around carbon tax, the degree of responsibility for climate change to be placed on large corporations and the proportion of public spending on roads versus public transport all remained contentious issues. One recommendation that did attract significant support was for new climate legislation, which would create an enhanced governance framework for climate action, establish an Oireachtas standing committee on climate change, replace the Climate Change Advisory Council with a Climate Action Council and mandate the setting of five-year carbon budgets and the creation of a target of net-zero economy-wide emissions by 2050.

In June 2019, the Government published a Climate Action Plan. This was stated as being built on the Report of the Joint Committee on Climate Action and is linked to 2030 Framework and the Paris Agreement. Running to over 150 pages, the Action Plan covered Government plans to significantly alter the governance structure around climate action in Ireland, but also set out some of the most detailed plans for many sectors of the economy, including agriculture, waste, the build environment, transport and electricity. As regards legislative change, it contained a commitment to publish a new Climate Action (Amendment) Bill in Quarter 1 of 2020. This would make provision for the adoption of carbon budgets a legal requirement, establish the Climate Action Council as a successor organisation to the Climate Change Advisory Council, and create a Long-Term Climate Strategy to match the period covered by the three five-year carbon budgets, which would replace the existing requirement for National Mitigation Plans. The new legislation was also to establish the 2050 target in law. These legislative proposals in the Climate Action Plan were swiftly published as the Draft Scheme of a Bill in January 2020, but two days after their publication the Dáil was dissolved and a general election called.

## Conclusion

Despite being bound by common international agreements and EU legislative measures, Ireland has consistently performed at the bottom of all Member State league tables with respect to climate action. A similar laxity with respect to environmental law generally has been identified (Torney and O'Gorman 2019), leading to Ireland making regular appearances before the European Court of Justice. Potentially, the distance to the 2020 deadline meant that consequences in terms of potential fines were considered by successive Governments to be so far distant that there was little political gain to be made from undertaking the significant changes to the economy needed to achieve the targets. This political inertia can also be seen in terms of Ireland's slowness in adopting domestic climate legislation, and, as this chapter has shown, even when the 2015 Act was passed, the law's ambition and design fell well short of good practice. The weakness and lateness of the domestic legislation meant there was nothing pushing climate action in the run up to the EU deadlines, deadlines which most politicians were already discounting.

Beyond the actual text of laws, it would appear that for much of the last decade, there was a generalised reluctance or refusal to actually implement climate policies, irrespective of their source, on the part of the politicians and the public service within the country. This is why it may be that 2019 will be seen as a watershed year in terms of Ireland's approach to climate policy. Budget 2020, published in October 2019, saw a €6 increase in the carbon tax and hypothecation of the additional revenue for climate action projects. The Climate Action Plan, discussed above, was not something that flowed from the Climate Action and Low Carbon Development Act 2015. It was a completely new policy initiative and represented a recognition by the Government that that legislation as originally drafted, as well as wider policy, was not sufficiently effective. It was also seen as a means of implementing the recommendations of the Citizens' Assembly, a deliberative process which had gained legitimacy through its role and that of its predecessor, the Constitutional Convention, in the Marriage Equality and Abortion referendum processes. The Climate Action (Amendment) Bill, consequent on the Climate Action Plan, actually addresses many of the weaknesses identified in section 'How Does the Climate Action and Low Carbon Act, 2015 Compare Internationally?'. As such, when the bill is passed, it can be thought of as 'climate action for slow learners'.

The reason 2019 saw this change in the political approach is directly linked to the enhanced public interest in climate change, driven by a greater media focus on environmental issues, the championing of climate change by the likes of David Attenborough and the unique phenomenon of Greta Thunberg. In Ireland, this change can be measured in terms of the strong showing of Green Party candidates in the May 2019 European and Local Elections, as well as the prominence given by members of the public to climate issues during those campaigns (RedC 2019). These factors directly influenced the directional change in legislation and policy that has occurred. As Ireland prepared for a general election in early 2020, the prominence that climate action takes in the political campaign and any subsequent negotiations on government formation will undoubtedly influence the extent to which the improved legislative structure will be mirrored by a more vigorous degree of policy implementation as the State works to achieve its all-important 2030 targets.

## Bibliography


Bauer, A., J. Feichtinger, and R. Steurer. 2012. The Governance of Climate Change Adaptation in 10 OECD Countries: Challenges and Approaches. *Journal of Environmental Policy & Planning* 14: 279–304.

Burck, J., U. Hagen, F. Marten, N. Höhne, and C. Bals. 2018. *The Climate Change Performance Index: Results 2019.* Bonn: Germanwatch.

Citizens' Assembly. 2018. *Third Report and Recommendations of the Citizens' Assembly: How the State can Make Ireland a Leader in Tackling Climate Change.* Available at: https://www.citizensassembly.ie/en/how-the-state-can-make-ireland-a-leader-in-tackling-climate-change/final-report-on-how-the-state-can-make-ireland-a-leader-in-tackling-climate-change/climate-change-report-final.pdf. Accessed 17 July 2020.

Climate Action Europe. 2018. 'Off Target: Ranking of EU Countries' Ambition and Progress in Fighting Climate Change. Available at: http://www.caneurope.org/docman/climate-energy-targets/3357-off-target-ranking-of-eu-countries-ambition-and-progress-in-fighting-climate-change/file. Accessed 17 July 2020.

Department of Climate Action, Communications and the Environment. 2017. *National Mitigation Plan*. Dublin.

———. 2018. *National Adaptation Framework*. Dublin.

Doran, P. 2013. The Climate Action and Low Carbon Development Bill (Draft Heads) – An Opinion: Part One. *Irish Planning and Environmental Law* 20 (3): 116–124.

Ecologic Institute. 2017. Paris Compatible Governance: Long-Term Policy Frameworks to Drive Transformational Change. Available at: https://www.ecologic.eu/sites/files/publication/2017/paris_compatible_governance_-_ecologic_institute_report_0.pdf. Accessed 17 July 2020.

Elkink, J.A., D.M. Farrell, T. Reidy, and J. Suiter. 2017. Understanding the 2015 Marriage Referendum in Ireland: Context, Campaign, and Conservative Ireland. *Irish Political Studies* 32 (3): 361–381.

Fankhauser, S., A. Averchenkova, and J. Finnegan. 2018. *10 years of the UK Climate Change Act*. The Grantham Research Institute on Climate Change and the Environment, The Centre for Climate Change Economics and Policy (CCCEP) LSE.

Gillard, R. 2016. Unravelling the United Kingdom's Climate Policy Consensus: The Power of Ideas, Discourse and Institutions. *Global Environmental Change* 40: 26–36.

Government of Ireland. 2019. *Climate Action Plan 2019*. Dublin.

Government Offices of Sweden; Ministry of the Environment and Energy. 2017. The Swedish Climate Policy Framework.

Irish Cooperative Organisation Society. 2016. ICOS Climate Change Meeting with Phil Hogan [Press Release], 12 June.

Joint Committee on Climate Action. 2019. *Climate Change: A Cross-Party Consensus for Action*. Dublin.

Kennedy, R. 2011. Climate Change Law and Policy in Ireland. *Finnish Environmental Law Review* 3: 66–83.

———. 2016. New Ideas or False Hopes?: International, European and Irish Climate Change Law and Policy After the Paris Agreement. *Irish Planning and Environmental Law Journal* 23 (3): 75–82.

Kingston, S., V. Heyvaert, and A. Cavoski. 2017. *European Environmental Law*. Cambridge: Cambridge University Press.

Linehan, C. 2012. Climate Change and the Planning and Development (Amendment) Act 2010. *Irish Planning and Environmental Law Journal* 19 (3): 115–119.

Lockwood, M. 2013. The Political Sustainability of Climate Policy: The Case of the UK Climate Change Act. *Global Environmental Change* 23: 1339–1348.
O'Connor, M. 2017. Gone with the Wind: The Uncertain Pursuit of Ireland's 2020 RES-E Target – An Overview. *Irish Planning and Environmental Law Journal* 24 (4): 148–157.
O'Sullivan, K. 2018. Budget's Failure to Increase Carbon Tax Is Shocking. *The Irish Times*, October 9.
RedC/RTE/TG4. 2019. *Exit Poll for the European Elections, Local Election and Divorce Referendum*, May 24.
Sustainable Energy Authority of Ireland. 2019. Energy in Ireland: 2019 Report, December, 2019.
Torney, D. 2017. If at First You Don't Succeed: The Development of Climate Change Legislation in Ireland. *Irish Political Studies* 32 (2): 247–267.
Torney, D., and R. O'Gorman. 2019. A Laggard in Good Times and Bad? The Limited Impact of EU Membership on Ireland's Climate Change and Environmental Policy. *Irish Political Studies* 32 (2): 247–267.
Transport & Environment. 2018. From Effort Sharing Decision to Climate Action Regulation: The Good, the Bad and the Ugly. Available at: https://www.transportenvironment.org/sites/te/files/publications/2018_03_ESR_CAR_final_report.pdf. Accessed 17 July 2020.

# The Party Politics of Climate Change in Ireland

*Conor Little*

## Introduction

On 10 May 2019, the Dáil (Ireland's lower house of parliament) declared a climate and biodiversity emergency. The declaration was contained in an amendment to a motion that noted the report of a cross-party parliamentary committee, the Joint Oireachtas Committee on Climate Action (JOCCA 2019). The local and European Parliament elections were two weeks away, climate change was becoming increasingly salient among the Irish public (European Commission 2020), and the UK Parliament had made a similar declaration earlier in the month. The amendment that would declare a climate emergency was tabled by Fianna Fáil, the largest opposition party. There were only six TDs (MPs) present when the amendment was discussed in the Dáil and there were no Fianna Fáil deputies among them, so the leader of the Green Party proposed the amendment and it was accepted without a vote. This episode is suggestive of a number

C. Little (✉)
Department of Politics and Public Administration, University of Limerick, Limerick, Ireland
e-mail: Conor.Little@ul.ie



D. Robbins et al. (eds.), *Ireland and the Climate Crisis*, Palgrave Studies in Media and Environmental Communication,
https://doi.org/10.1007/978-3-030-47587-1_6

of features of climate politics in Ireland that are explored further in this chapter: the low salience of climate policy among the parties, a tendency towards party political consensus, and a willingness on the part of politicians to respond—at least symbolically—to voters' concerns.

Normatively, parties are expected to represent voters' preferences. They are also expected to solve policy problems, including problems like climate change, which have effects that extend beyond local and national boundaries and beyond the electoral cycle. The responsibility to address these problems presents tensions with parties' incentives to deliver local and national benefits within the electoral cycle. Where climate policy is not a priority for voters—as has typically been the case in Ireland—this presents further tensions between solving this policy problem and seeking votes.

Empirically, parties and their representatives matter for climate policy and politics. Most significantly, they influence government policy outputs, such as taxes, subsidies, and regulations, and ultimately these policies influence greenhouse gas (GHG) emissions (Harrison 2010; Jensen and Spoon 2011; Birchall 2014; Carter and Jacobs 2014). Parties also influence the behaviour and attitudes of other actors, including members of the public (Guber 2013; Sohlberg 2016; Carmichael and Brulle 2017).

Parties' behaviour and preferences on climate policy are influenced, in turn, by a variety of factors. Existing research finds that parties respond to public opinion and party competition on climate policy and to the politics of coalition. Their responses are also shaped by intra-party factors, including their pre-existing policy commitments on economic policy and other issues. Left-of-centre parties often—although not always—afford greater priority to the problem of climate change than right-of-centre parties. And overall, climate change has not been a top priority for most parties most of the time in industrialised democracies (Carter 2013; Båtstrand 2014, 2015; Marcinkiewicz and Tosun 2015; Carter et al. 2018; Farstad 2018, 2019; Ladrech and Little 2019).

This chapter provides an introduction to how parties and their representatives have responded to and influenced climate politics and policy in Ireland. It does so, first, with a review of research on the party politics of climate change in Ireland; second, with an overview of how the parties' responses have developed on some important climate policy issues since the late 1990s; and, third, by identifying ways in which the case of Ireland can contribute to knowledge on the party politics of climate change more generally. In doing so, it highlights several ways in which Ireland is substantively and theoretically an interesting case.

## Research on the Party Politics of Climate Change in Ireland

There is a small number of studies with an explicit theoretical focus on the party politics of climate policy and a substantial empirical focus on Ireland.[1] Drawing on evidence from the party politics of climate change in Denmark and Ireland, Ladrech and Little (2019) show that there are important commonalities in how political parties' climate policy preferences are shaped by their pursuit of political goals but that differences in parties' opportunity structures in the two countries have produced quite different responses and structures of party competition on climate change. As a result, climate policy preferences among Ireland's parties are, on average, less 'green' than those of parties in Denmark. This is partly because of demand-side factors, especially low public concern with climate change in Ireland, and partly due to the lack of large or numerous ecologically oriented parties.

Other studies of climate politics examine the inner workings of parties. Dave Robbins (2019: f.104) has looked inside the Green Party during its spell in government from 2007 to 2011, finding diverse approaches in how the two Green senior ministers and their advisers sought to influence the media agenda on climate change. Other research has examined the role and effects of 'intra-party policy entrepreneurs' who have pursued significant change in their party's climate policies. Typically, these individuals have not succeeded in Ireland, as their climate policy-seeking behaviour has fallen foul of their parties' office-seeking incentives. Yet, on the rare occasions when office-seeking incentives have aligned with greener climate policies, they have driven greener climate policy positions, greater salience for climate policy, and have influenced government policy (Little 2017a).

Themes that emerge from these studies include the low priority of climate policy for most Irish parties and the relatively few policy differences between them. These observations are further supported by comparing Irish parties to parties in other countries (Carter et al. 2018).

While research on the Irish case that explicitly focuses on party politics has been rare, parties feature empirically in a broader set of studies on key climate policies in Ireland. Coghlan (2007) provided an early account of how policymakers responded to the emergence of climate change as an issue up to 2004. In particular, he focused on the disjuncture between international commitments made in the late 1990s and domestic policy-making, highlighting the role of intra-party and intra-coalition dynamics.

Coghlan's (2007) study is, in large part, focused on the first attempt to see a *carbon tax* introduced in Ireland, until its abandonment in 2004. Dunne et al. (2013) examine the second, successful, attempt to introduce a carbon tax, and they underline the important role of the Green Party in government for the introduction of the tax in late 2009, in conjunction with economic events that were upending the Irish economy. Andersen (2019) examines Ireland's carbon tax in comparative perspective, reserving a central place for the diversity among parties facilitated by proportional representation in his explanation of the adoption of carbon taxes in small states like Ireland.

Parties are also central to accounts of the pathway to *framework climate legislation* in Ireland, which culminated in the enactment of the Climate Action and Low Carbon Development Act under a Fine Gael/Labour coalition in 2015 (see chapter "Climate Law in Ireland: EU and National Dimensions" by O'Gorman, this volume). Torney (2017, 2019) shows how domestic politics constrained the diffusion of climate legislation from the UK, effectively watering down the model presented by the UK Climate Change Act. While economic interest groups worked actively to achieve this end, Torney's case studies also identify parties and their representatives as central actors in developing Ireland's weaker legislation. Likewise, Wagner and Ylä-Anttila (2018), taking an advocacy coalitions approach and using network analysis of unique data on Ireland's climate policy community—including political parties—show that economic interest groups' and government parties' preferences were reflected in the 2015 Act.

The role of parties in climate politics in Ireland, as elsewhere, extends beyond influencing government policy. Like Robbins (2019), Wagner and Payne (2017) focus on *media coverage* of climate change, highlighting the central role of parties in discourse coalitions, where they have contributed to the dominant ecological modernisation perspective (see also chapter "Climate Change and the Media" by Robbins, this volume). In addition, Paul Cunningham's (2008) book, based on a series of interviews, provides insights into the perspectives of two government ministers who had responsibility for climate policy (Noel Dempsey and John Gormley).

In common with comparative research on parties and climate policy, these studies of climate policy and politics in Ireland attest to the significance of political parties and their representatives. This is further supported by a range of observations by researchers of climate politics in

Ireland: Ó'Broin (2015: 125–130, 135) noted the significance of climate change failing to become part of mainstream political discourse, Brennan and Curtin (2008: 135) highlighted the need for 'greater political ambition' on climate change, and Kirby (2013: 88) underlined the significance of 'political mobilization' for decarbonisation. In summary, the importance of parties and their representatives for climate politics and policy in Ireland is not in doubt, but the study of parties' preferences, behaviour, and influence has only sometimes been a central theoretical concern.

## The Party Politics of Climate Change in Ireland Since the Late 1990s

This section provides an overview of the party politics of climate change mitigation policy in Ireland from the late 1990s until mid-2019. It encompasses the period of Fianna Fáil/Progressive Democrat coalition from 1997 to 2007; the addition of the Greens to that coalition from 2007; and Fine Gael-led governments from 2011 onwards—in coalition with Labour until 2016 and with some non-party (Independent) TDs from 2016 on, supported by a confidence-and-supply agreement with Fianna Fáil. It focuses on the cross-sectoral issues of carbon taxation and framework climate legislation and on policy issues in Ireland's largest emissions sectors: agriculture, transport, and electricity generation.

Climate policy has typically been of low salience among the public and the parties, with two partial exceptions. One episode of raised salience occurred in 2006–2007 and another in 2019. In both instances, these peaks in salience coincided with strong economic growth and low levels of unemployment. Each was partly driven by international events (e.g. the publication of reports by the Intergovernmental Panel on Climate Change and mobilisation and policy responses in other countries) and by political competition with the Greens—approaching the 2007 general election, when they seemed likely to gain seats and after their surprise electoral success at the 2019 local and European elections. In 2019, increased media attention to climate change impacts, increased attention from policymakers in Ireland (e.g. the JOCCA), and new movements like Fridays for Future Ireland and Extinction Rebellion Ireland made climate change more salient. Increased concern among the public elicited responses from some of the parties and candidates in 2007 (Little 2017a, b; Ladrech and Little 2019) and from all parties, to varying extents, in 2019. Yet, in both instances, the level of public concern about climate change was dwarfed by concerns about other issues (crime, health, and the cost of living in 2006–2007; health and housing in 2019), so climate change remained a lower-order issue. Between these peaks, and especially from 2008 until 2014, the effects of a severe economic crisis depressed public and political concern about climate change (European Commission 2020).

During the period covered, the Greens consistently took distinctive positions on climate policy issues, but they were also consistently small (never holding more than 4% of Dáil seats) and there was a lack of disagreement on many climate policy issues among the other parties. At the other end of the spectrum, explicit climate science scepticism has been almost absent from party politics. Climate change adaptation, as such, has not yet become a party political issue, although local developments and incidents (e.g. related to flooding) occasionally impinge on representative politics.

### *Carbon Tax*

Carbon tax has been on the Irish policy agenda—intermittently—for almost as long as climate change itself. An initial attempt to introduce a carbon tax was driven by a Fianna Fáil Minister for the Environment, Noel Dempsey, who made it a central component of Ireland's first National Climate Change Strategy (NCCS) (NCCS 2000: 3). Its abandonment was the result of a coalition of Dempsey's party colleagues including the Minister for Finance, Charlie McCreevey, the Progressive Democrats, and industry (Coghlan 2007: 146–151). Neither Fine Gael nor Labour supported the introduction of a carbon tax—although there were differences of opinion on the issue within each party—while the Greens did support it.

The carbon tax was eventually introduced by Fianna Fáil and the Greens in late 2009 under significant fiscal duress (Dunne et al. 2013). Industry changed its position on the tax, and Fine Gael and Labour increased the rate from €15 to €20 per tonne in 2013. In 2018 and 2019, increasing the carbon tax was the most high-profile issue in the Joint Oireachtas Committee on Climate Action's (JOCCA) deliberations and it generated the most significant level of disagreement between the parties. Most parties on the committee agreed that the rate should increase to €80 per tonne by 2030, following the advice of the Climate Change Advisory Council (JOCCA 2019). However, the third-largest party at the time, Sinn Féin, and the left-wing Solidarity-People Before Profit party opposed carbon tax increases (PBP 2019; Sinn Féin 2019). In late 2019, the Fine Gael-led coalition increased the rate of carbon tax to €26 per tonne.

### *Climate Change Legislation*

The Greens and Labour played entrepreneurial roles in relation to proposals for framework climate change legislation in the years before the enactment of the Climate Action and Low Carbon Development Act 2015. While the Greens were in government (2007–2011), Labour introduced Private Members' Bills that included emissions targets and carbon budget mechanisms and, in October 2009 and 2010, the cross-party Joint Oireachtas Committee on Climate Change and Energy Security published proposals for a climate law, also including a 2050 target and carbon budgets. Within government, the Greens secured a commitment to a climate law including emissions reductions targets. It published a framework for the bill in late 2009 and a draft law one year later that contained annual targets to 2020 and targets for 2030 and 2050 (Torney 2017). There was strong resistance from within Fianna Fáil to the Green Party's legislation, albeit the Fianna Fáil leadership insisted it was necessary in order to sustain the coalition (interview with Green Party adviser, July 2015).

From 2011, Fine Gael's Minister for the Environment Phil Hogan (succeeded as Minister by Labour's Alan Kelly in 2014) had responsibility for progressing the government's commitment to climate legislation. When a draft bill was published in October 2013, it had jettisoned all quantitative targets, although later set a non-legislative target of an 80% reduction in GHG emissions by 2050 across non-agricultural sectors and 'an approach to carbon neutrality in the agriculture and land-use sector, including forestry, which does not compromise capacity for sustainable food production' (Government of Ireland 2014).

The government made some concessions, including a commitment to 'climate justice' in the legislation on foot of pressure from a leftist non-party TD (Maureen O'Sullivan) and by guaranteeing the independence of the Climate Change Advisory Council, which was established by the legislation, but saw its preferences enacted on the absence of targets and carbon budgets (Torney 2017: 252–260; Wagner and Ylä-Anttila 2018). The Act received cross-party support, and there was effective consensus on the key issue of targets between Fianna Fáil and Fine Gael (interview with non-governmental organisation [NGO] representative, June 2016). In 2019, climate change legislation became the subject of a new cross-party consensus: the members of the cross-party committee agreed that there should be new legislation enacted that would include targets and carbon budgets (JOCCA 2019; Sinn Féin 2019).

### *Agriculture*

Agriculture is Ireland's largest emissions sector, accounting for about one-third of GHG emissions. Since the late 1990s, two phases in agricultural climate policy, both of which reflect considerable cross-party consensus, can be identified. Before the economic crisis and as the rest of the economy grew rapidly, agricultural climate policy was of very low salience; indeed, the connection between agriculture and climate change was rarely made. The Fianna Fáil-led government envisaged a decline in cattle and sheep herd numbers of 5% contributing to emissions reductions in the first NCCS (2000); while this was not explicitly included in the revised Strategy (NCCS 2007), that document continued to envisage decline in agricultural emissions.

With the publication of the *Food Harvest 2020* policy in 2010, agriculture was reframed as an engine of export-led economic growth, with plans to significantly increase Ireland's national herd. Changed economic circumstances and effective lobbying led to a consensus supportive of the expansion of emissions-intensive agriculture among parties with any significant electoral ambitions in rural areas, that is, Fine Gael, Fianna Fáil, Sinn Féin, and Labour. At the Paris COP in December 2015, Taoiseach Enda Kenny told an international audience that Ireland would 'play its part' in addressing climate change, but briefed Irish journalists that agriculture would not be compromised by climate policy. The opposition parties' responses were muted. In a televised debate among the leaders of the four main parties in early 2016, none of the main party leaders, when questioned about the conflict between climate policy and agriculture policy, conceded that reduced ambition in meat and dairy production was necessary to achieve Ireland's climate policy objectives. This consensus was also reflected in the 'confidence and supply' agreement struck between Fine Gael and Fianna Fáil in mid-2016, which recommitted the parties to ambitious herd expansion, but did not mention climate policy.

The cross-party consensus on agriculture has evolved a little since 2016, with the 2019 report of the JOCCA reflecting an acceptance—in principle—of the need to examine diversification and land-use change in agriculture (JOCCA 2019: Ch. 8). The main parties remained opposed to extending a carbon tax to the agriculture sector.

### *Electricity*

Electricity generation and transmission is another issue around which there has been a broad but shifting consensus. The ambition of increasing electricity generation from renewable sources has attracted broad cross-party support. This was especially the case in the 2000s. However, as infrastructure, including wind turbines and pylons for electricity transmission, was planned and deployed, it provoked mobilisation by local campaigns of opposition from around 2010, and the content of that inter-party consensus changed substantially. By 2016, Fianna Fáil, Fine Gael, and Labour were indicating that they wanted the development of onshore wind energy slowed down or stopped (Augustenborg 2016).

On the future of Ireland's fossil fuel-burning electricity generation plants, including its coal-burning plant at Moneypoint and several peat-burning plants, electoral politics in rural areas has played a role in their opening (including in the 2000s), maintenance, and continued use (e.g. Cunningham 2008: 107). By mid-2019, the members of JOCCA (2019: 58) had expressed a preference for phasing out peat and coal 'as soon as it is technically feasible' and recommended continuously examining possibilities for achieving this, although they did not set a deadline.

### *Transport*

The absence of financially and socially acceptable policy solutions in transport climate policy is reflected in one Transport Minister blaming the public for rising emissions in the late 2000s: 'The department hasn't gone the wrong way – it's the public' (Dempsey quoted in Cunningham 2008: 108). These constraints have contributed to a consensus that has been largely in favour of the existing transport system, although the Greens and, at times, Labour (e.g. in its 2002 general election manifesto) have clearly prioritised public transport. JOCCA (2019) agreed that 10% of transport programmes should be allocated to cycling but rejected long-standing Green Party policies of spending twice as much on public transport as on roads and reducing speed limits by 10 km/h (McGee 2019).

With some exceptions, the parties have tended to support, or not to oppose, expanding aviation infrastructure, especially regional airports; like agricultural expansion, support for regional air connectivity is associated with the geographical breadth of parties' electoral ambitions. For instance, Labour renewed its support for regional airports in its 2011 general election manifesto when it aimed to win seats in rural areas. The Greens have been the main exception to the consensus on aviation, supporting the inclusion of aviation in the European Union (EU) Emissions Trading System (ETS) in 2016, for example. On rare occasions where government policy has deviated from the *status quo*, it has been seized upon as a matter for political advantage: in 2009, the Fianna Fáil/Green government introduced a small tax on air travel as part of their revenue-raising efforts during the economic crisis; Fine Gael campaigned against the tax and reversed it in 2014.

Overall, the strong tendency towards relative consensus on climate policy issues is evident across policy areas and has survived political and economic volatility. Where there are divisions between parties, they have reflected division between centre-right and centre-left, with the latter pushing for greener climate policies, but also an emerging division between parties further to the left, which most strongly oppose increased costs for individuals (e.g. on the carbon tax), favouring imposing costs on businesses and the other parties. Although the Greens' ownership of the climate policy issue was challenged by Labour during the 2007–2011 period, they remain the effective 'issue-owner' on climate policy and they also have taken distinctive climate policy positions most frequently.

## Using the Case of Ireland to Understand the Party Politics of Climate Change

This section identifies some ways in which the case of Ireland can contribute to debates on the party politics of climate change and some gaps in our empirical knowledge about the party politics of climate policy in Ireland. I highlight six potential contributions to the broader literature.

First, as noted above, Irish politics has been characterised by an unusual degree of consensus. The content of Ireland's policy consensuses on climate policy differs between issues and sectors, with a variety of implications for climate change mitigation policies. They have not always—or even frequently—been 'green' consensuses, in support of policies that would reduce GHG emissions or enhance sinks. Existing studies have focused on how party political consensus has facilitated policy that mitigates emissions (Carter and Jacobs 2014; Eikeland and Inderberg 2016) and on the suggestion that polarisation is an obstacle to effective climate policy (Christoff and Eckersley 2011: 442–443; Farstad 2019). The case of Ireland illustrates the simple proposition that the effects of policy consensus depend on the content of that consensus. On some sectors and issues, Ireland illustrates patterns of consensus underlying climate policy laggardship (Little 2017b).

Second, existing literature has focused on how consensus on climate policy is associated with valence competition, in which parties try to outdo one another on credibility or perceived competence (e.g. Carter and Jacobs 2014). While consensus is necessary for valence competition, the case of Ireland suggests that it is not sufficient. Only on a couple of occasions has climate mitigation policy become a moderately salient issue for political competition between the main parties, and even in these instances, party competition on climate policy has not been as intense as it has been in other contexts, such as the late 2000s in the UK (Carter and Jacobs 2014) and Denmark (Ladrech and Little 2019).

Third, the case of Ireland also adds evidence in support of the proposition that parties' preferences on climate policy are bound up with their preferences in other policy domains (see Carter 2013; Båtstrand 2014, 2015; Carter et al. 2018; Farstad 2018; Ladrech and Little 2019). The low level of inter-party differences on economic and social issues—which are in turn underpinned by factors such as the historical structure of the party system and aspects of the Irish electoral system—has underpinned the low level of inter-party differences on climate policy.

Fourth, by virtue of its experience of economic boom, bust, and recovery, Ireland offers an extreme case (Seawright and Gerring 2008: 301–302) of economic change, which can be used to probe the effects of economic change on the party politics of climate change.

Fifth, Ireland can function as a 'least-likely' case for partisan patterns of climate policy due to limited differences among the parties: it is plausible that if partisan climate policy obtains in Ireland—and there is already some evidence that it does, at least among junior coalition partners such as the Progressive Democrats, the Greens, and Labour—it is likely to obtain elsewhere.

Sixth, the literature on climate policy is missing a systematic theorisation and analysis of the party politics of climate change adaptation, that is, reducing vulnerability to climate change (Javeline 2014: 424, 427). Ireland and other cases can contribute to this. Adaptation is a major and growing issue, which increasingly involves the distribution of resources, as well as regulation. While significant elements of mitigation may be driven by private actors within a conducive regulatory framework, adaptation may bear more directly on government spending, thus correlating even more strongly than mitigation with party preferences on spending. The case of Ireland, with limited diversity in the party system on public spending preferences, may help to show how a small neoliberal state with a history of convergent partisan preferences responds to demands for adaptation.

In addition, there are empirical gaps in the study of the party politics of climate change in Ireland. Some of the most obvious include studies of Irish voters' attitudes and behaviours on climate change; this may become of increased interest as variation in concern about climate change (over time and between voters) increases (European Commission 2020). There is also relatively little research that focuses on representative politics in Ireland at local level or European level, although there is some work on the local politics of climate change in Ireland upon which such research could build (McGloughlin and Sweeney 2011; Dekker 2019).

Finally, there is a notable lack of sector-specific studies on, for example, the party politics of agricultural climate policy. While cross-sectoral issues like the carbon tax and climate legislation have consumed much political and academic attention to climate policy, much of the politics of climate change takes place at sectoral level and sectors and their constellations of interest groups differ considerably. Sectoral studies of climate politics will build on existing work on governance and politics in these policy domains (e.g. Adshead 1996) and can be linked to broader themes: studies of agriculture, for instance, might shed light on the politics of emissions-intensive, export-oriented industries.

## Conclusion

Parties matter for climate policy and politics. The party politics of climate change in Ireland has been characterised by low salience, with some variation over time and relatively low levels of inter-party disagreement. Peculiarities of Irish party politics that pre-date climate policy—including its durable centre-right consensus on social and economic policies—have underpinned this tendency towards consensus on climate policy issues. For this to become a 'greener', ambitious climate policy consensus, party positions and, perhaps, the salience of climate policy, would need to change.

For researchers, there is considerable further ground to be covered in understanding the past development and possible futures of the climate politics in Ireland, and understanding political parties will be central to that task. Country studies and comparative studies with a substantial focus on Ireland have significant potential to inform the broader literature on the party politics of climate change. Empirically, there is much common ground between studies of the party politics of climate change and studies of policy processes and non-party actors, such as interest groups and officials, which mean that these bodies of literature will continue to inform one another.

Ireland's climate politics may not be a low-salience consensus forevermore: while voters have rarely troubled the Irish parties about climate change in large numbers, it is likely to become more salient, either cyclically or as a secular trend. However, climate policy will remain in competition with other issues on a crowded policy agenda. This was illustrated at the general election of February 2020, when the proportion of the public that viewed climate change as a major problem for Ireland, while high in historical terms, was much lower than the proportion concerned about other policy problems, especially housing and healthcare, and was lower than in the average EU member state (European Commission 2020; see also Collins 2020). Climate change has some way to go before it becomes a top-tier issue in Irish politics.

Ireland's history of policy consensus suggests that, were a 'green' climate policy consensus to be forged, party politics in Ireland has the potential to sustain such a consensus. The development of the report of the JOCCA (2019) shows how parties can build on that potential to actively develop agreement with input from experts and stakeholders, while at the same time building capacity among parliamentarians to engage with climate policy.

Nonetheless, the development of any such green consensus faces challenges. The results of the general election of February 2020 suggested increased polarisation on climate policy. On carbon tax, parties that did not support the JOCCA recommendation that the rate be progressively increased to €80 per tonne by 2030—Sinn Féin, Solidarity-People Before Profit, and the Social Democrats—won 48 seats (of 160, up from 30 when the election was called), with Sinn Féin winning a plurality of the vote and 37 seats to become the second-largest party in the Dáil. Many of the 20 non-party TDs that were elected would share their position on carbon tax increases. Fianna Fáil (38 seats), Fine Gael (35 seats), and Labour (5 seats) supported the JOCCA recommendation in their election manifestos, while the Green Party, which won 12 seats—their best general election result to date—advocated a sharper upward trajectory towards €100 per tonne in 2030.

Climate policy and particularly policies that impose costs or challenge existing behaviour generates opposition, so increased salience and proposals for policy change are likely to go hand-in-hand with pressures towards disagreement and polarisation. The future of climate politics may be more positional as governments move from lower-cost to higher-cost emissions abatement options. Parties are likely to be at the frontline in the communication of climate policies for societal acceptance and may need to develop new ways of communicating these policies (see, e.g. FitzGerald 2019).

With regard to JOCCA (2019), it must also be borne in mind that Oireachtas committees' recommendations are not always the most enduring indicator of parties' preferences. In 2010, a predecessor of JOCCA, the Joint Committee on Climate Change and Energy Security, proposed climate legislation that included a legislative target and carbon budgets; within a short time, some of the parties that represented on that committee had proposed weaker legislation with no legislative targets and no carbon budgets.

## Note

1. This chapter focuses on the party politics of climate change, but of course this topic has strong links with the politics of environmental protection more generally (Carter 2018), and research on environmental politics in Ireland provides a basis for research on climate politics (e.g. Taylor 1998; Flynn 2006; Laffan and O'Mahony 2008; O'Mahony and Keohane 2011; Connaughton 2019).

## Bibliography

Adshead, M. 1996. Beyond Clientelism: Agricultural Networks in Ireland and the EU. *West European Politics* 19 (3): 583–608.

Andersen, M.S. 2019. The Politics of Carbon Taxation: How Varieties of Policy Style Matter. *Environmental Politics* 28 (6): 1084–1104.

Augustenborg, C. 2016. Election 2016 Climate Manifesto Analysis. *Cara Augustenborg*. Available from: http://www.caraaugustenborg.com/1/post/2016/02/election-2016-climate-manifesto-analysis.html. Accessed 29 Aug 2016.

Båtstrand, S. 2014. Giving Content to New Politics. From Broad Hypothesis to Empirical Analysis Using Norwegian Manifesto Data on Climate Change. *Party Politics* 20 (6): 930–939.

———. 2015. More than Markets: A Comparative Study of Nine Conservative Parties on Climate Change. *Politics & Policy* 43 (4): 538–561.

Birchall, S.J. 2014. Termination Theory and National Climate Change Mitigation Programs: The Case of New Zealand. *Review of Policy Research* 31 (1): 38–59.

Brennan, P., and J. Curtin. 2008. *Climate Change Challenge- Strategic Issues, Options and Implications for Ireland*. Dublin: IIEA.

Carmichael, J.T., and R.J. Brulle. 2017. Elite Cues, Media Coverage, and Public Concern: An Integrated Path Analysis of Public Opinion on Climate Change, 2001–2013. *Environmental Politics* 26 (2): 232–252.

Carter, N. 2013. Greening the Mainstream: Party Politics and the Environment. *Environmental Politics* 22 (1): 73–94.

———. 2018. *The Politics of the Environment: Ideas, Activism, Policy*. 3rd ed. Cambridge: CUP.

Carter, N., and M. Jacobs. 2014. Explaining Radical Policy Change: The Case of Climate Change and Energy Policy Under the British Labour Government 2006–10. *Public Administration* 92 (1): 125–141.

Carter, N., R. Ladrech, C. Little, and V. Tsagkroni. 2018. Political Parties and Climate Policy: A New Approach to Measuring Parties' Climate Policy Preferences. *Party Politics* 24 (6): 731–742.

Christoff, P., and R. Eckersley. 2011. Comparing State Responses. In *Oxford Handbook of Climate Change and Society*, ed. J.S. Dryzek, R.B. Norgaard, and D. Schlosberg, 431–448. Oxford: OUP.

Coghlan, O. 2007. Irish Climate-Change Policy from Kyoto to the Carbon Tax: A Two-Level Game Analysis of the Interplay of Knowledge and Power. *Irish Studies in International Affairs* 18 (1): 131–153.

Collins, S. 2020. Detailed Election 2020 Exit Poll Results: How Voters Answered 15 Questions. *The Irish Times*. Available from: https://www.irishtimes.com/news/politics/detailed-election-2020-exit-poll-results-how-voters-answered-15-questions-1.4167016. Accessed 9 Feb 2020.

Connaughton, B. 2019. *The Implementation of Environmental Policy in Ireland: Lessons from Translating EU Directives into Action*. Manchester: MUP.

Cunningham, P. 2008. *Ireland's Burning*. Dublin: Poolbeg.

Dekker, S. 2019. *Cities Leading Climate Action: Urban Policy and Planning*. New York: Routledge.

Dunne, L., D. Joyce, and F. Convery. 2013. Ireland's Carbon Tax and the Fiscal Crisis. *OECD Environment Working Papers* (59).

Eikeland, P.O., and T.H.J. Inderberg. 2016. Energy System Transformation and Long-Term Interest Constellations in Denmark: Can Agency Beat Structure? *Energy Research & Social Science* 11: 164–173.

European Commission. 2020. Standard Eurobarometer Surveys. *PublicOpinion – European Commission*. Available from: https://ec.europa.eu/commfrontoffice/publicopinion/index.cfm/General/index. Accessed 10 Feb 2020.

Farstad, F.M. 2018. What Explains Variation in Parties' Climate Change Salience? *Party Politics* 24 (6): 698–707.

———. 2019. Does Size Matter? Comparing the Party Politics of Climate Change in Australia and Norway. *Environmental Politics* 28 (6): 997–1016.

FitzGerald, D.C. 2019. *The Framing of Climate Action in Ireland: Strategic Considerations*, NESC Secretariat Papers, No.17. Dublin: NESC.

Flynn, B. 2006. *The Blame Game: Rethinking Ireland's Sustainable Development and Environmental Performance*. Dublin: Irish Academic Press.

Government of Ireland. 2014. *Climate Action and Low-Carbon Development: National Policy Position Ireland*. Dublin: Government of Ireland.

Guber, D.L. 2013. A Cooling Climate for Change? Party Polarization and the Politics of Global Warming. *American Behavioral Scientist* 57 (1): 93–115.

Harrison, K. 2010. The Comparative Politics of Carbon Taxation. *Annual Review of Law and Social Science* 6 (1): 507–529.

Javeline, D. 2014. The Most Important Topic Political Scientists Are Not Studying: Adapting to Climate Change. *Perspectives on Politics* 12: 420–434.

Jensen, C.B., and J.-J. Spoon. 2011. Testing the 'Party Matters' Thesis: Explaining Progress Towards Kyoto Protocol Targets. *Political Studies* 59 (1): 99–115.

JOCCA. 2019. *Report of the Joint Committee on Climate Action. Climate Change: A Cross-Party Consensus for Action*. Dublin: Houses of the Oireachtas.

Kirby, P. 2013. Policy Optimism: NESC, Climate Change and Achieving Decarbonisation. *Administration* 61 (2): 75–90.

Ladrech, R., and C. Little. 2019. Drivers of Political Parties' Climate Policy Preferences: Lessons from Denmark and Ireland. *Environmental Politics* 28 (6): 1017–1038.

Laffan, B., and J. O'Mahony. 2008. 'Bringing Politics Back In'. Domestic Conflict and the Negotiated Implementation of EU – Nature Conservation Legislation in Ireland. *Journal of Environmental Policy & Planning* 10 (2): 175.

Little, C. 2017a. Intra-Party Policy Entrepreneurship and Party Goals: The Case of Political Parties' Climate Policy Preferences in Ireland. *Irish Political Studies* 32 (2): 199–223.

———. 2017b. Portrait of a Laggard? Environmental Politics and the Irish General Election of February 2016. *Environmental Politics* 26 (1): 183–188.

Marcinkiewicz, K., and J. Tosun. 2015. Contesting Climate Change: Mapping the Political Debate in Poland. *East European Politics* 31 (2): 187–207.

McGee, H. 2019. Climate Change: Plotting an Action Plan for Ireland. *The Irish Times*, March 30.

McGloughlin, J.S., and J. Sweeney. 2011. Multi-Level Climate Policies in Ireland. *Irish Geography* 44 (1): 137–150.

NCCS. 2000. *National Climate Change Strategy*. Dublin: Government of Ireland.

———. 2007. *National Climate Change Strategy 2007–2012*. Dublin: Government of Ireland.

Ó'Broin, D. 2015. Conclusion: Politico-Cultural Change and Institutional Re-design. In *Adapting to Climate Change: Governance Challenges*, ed. D. Ó'Broin and P. Kirby, 119–137. Glasnevin Pub: Dublin.

O'Mahony, P., and K. Keohane. 2011. *Irish Environmental Politics After the Communicative Turn*. Manchester: Manchester University Press.

PBP. 2019. *Minority Statement on the Report of the Joint Committee on Climate Action*. Dublin: People Before Profit.

Robbins, D. 2019. *Climate Change, Politics and the Press in Ireland*. London: Routledge/Taylor & Francis Group.

Seawright, J., and J. Gerring. 2008. Case Selection Techniques in Case Study Research. *Political Research Quarterly* 61 (2): 294–308.

Sinn Féin. 2019. *Pathways to Equitable and Sustainable Climate Change in Ireland. Minority Report*. Dublin: Sinn Féin.

Sohlberg, J. 2016. The Effect of Elite Polarization: A Comparative Perspective on How Party Elites Influence Attitudes and Behavior on Climate Change in the European Union. *Sustainability* 9 (1): 39.

Taylor, G. 1998. Conserving the Emerald Tiger: The Politics of Environmental Regulation in Ireland. *Environmental Politics* 7 (4): 53–74.

Torney, D. 2017. If at First You Don't Succeed: The Development of Climate Change Legislation in Ireland. *Irish Political Studies* 32 (2): 247–267.

———. 2019. Climate Laws in Small European States: Symbolic Legislation and Limits of Diffusion in Ireland and Finland. *Environmental Politics* 28 (6): 1124–1144.

Wagner, P., and D. Payne. 2017. Trends, Frames and Discourse Networks: Analysing the Coverage of Climate Change in Irish Newspapers. *Irish Journal of Sociology* 25 (1): 5–28.

Wagner, P., and T. Ylä-Anttila. 2018. Who Got Their Way? Advocacy Coalitions and the Irish Climate Change Law. *Environmental Politics* 27 (5): 872–891.

# Responding to Climate Change: The Role of Local Government in Ireland

*Sabrina Dekker*

## Introduction

National policy documents highlight the important role of local authorities in responding to climate change, including the *National Mitigation Plan* (NMP) (Department of Communications, Climate Action and Environment 2017), *National Adaptation Framework* (NAF) (Department of Communications, Climate Action and Environment 2018b), the Joint Oireachtas Report *Climate Change: A Cross-Party Consensus For Action* (Houses of the Oireachtas 2019), and the 2019 *Climate Action Plan* (Department of Communications, Climate Action and Environment 2019). National Government emphasises the role of Irish local authorities as being facilitators and supporters of community-led initiatives. As public sector bodies, local authorities are expected to contribute to national targets for carbon emissions and lead by example with the uptake of energy efficiency measures across their buildings and more broadly in their service delivery. Furthermore, local authorities are mandated to develop climate

S. Dekker (✉)
Dublin City University, Dublin, Ireland



D. Robbins et al. (eds.), *Ireland and the Climate Crisis*, Palgrave Studies in Media and Environmental Communication,
https://doi.org/10.1007/978-3-030-47587-1_7

adaptation plans that have to incorporate sectoral adaptation plans. These are not unique expectations, as globally it is well recognised that local governments are well positioned to respond to climate change. However, not all local governments are created equal and nowhere is this more evident than in Ireland, where local authorities are restricted in their operations and service delivery. Irish local authorities, with their limited remits and financial and legal capacity, nonetheless have the potential to demonstrate how, despite the challenges, climate action is possible with limited resources, finance and legal capacity.

This chapter investigates how Irish local authorities are making progress on climate action working within the constraints they face. The data underlying this chapter comes from a review of existing policies, interviews with key stakeholders and the author's first-hand experience with developing the four Dublin local authorities' climate action plans. The chapter begins with a discussion on the institutional structure within which Irish local authorities operate. Critical to this is the Irish planning context and public consultation, which is a feature of policymaking in Ireland, yet presents challenges for climate action. In order to highlight the varying differences and opportunities inherent in responding to climate change for local government in Ireland, the chapter discusses the responses employed by cities globally. Finally, the chapter discusses the Irish approach to climate action using a case study of Dublin's climate action plan development and concludes with recommendations for strengthening climate action at local authority level in Ireland.

## Institutional Structure

The *Climate Action and Low Carbon Development Act 2015* was passed by the Oireachtas in December 2015 (Department of Environment Community and Local Government 2015). The Act sets out how Ireland is to respond to climate change (see also chapter "Climate Law in Ireland: EU and National Dimensions" by O'Gorman, this volume). There is a call for an all-of-government approach in the Act, as responding to climate change involves all sectors and departments alongside semi-state bodies. However, the role of local authorities in the Act is not as robust or significant as it could be. This is reflected in the *National Mitigation Plan* (NMP) (Department of Communications, Climate Action and Environment 2017) and the *National Adaptation Framework* (NAF) (Department of Communications, Climate Action and Environment

2018b). The NMP focuses on mitigation action on a sectoral basis. Similarly, the NAF applies a sectoral approach to adaptation. The latter, though, acknowledges the role of local authorities, but this is nominal as will be discussed. Arguably, a sectoral approach runs counter to the collaboration implied in the Act. The sectoral approach has not changed significantly with the 2019 *Climate Action Plan* (Department of Communications, Climate Action and Environment 2019) that was developed to address the weaknesses of the NMP. Local authorities are given responsibility for a few future-based actions, but for the most part continue to be expected to respond to the everyday needs of National Government departments.

Local authorities are primarily dependent on the ability of the Department of Housing, Planning and Local Government (DHPLG) to acquire funds from the Department of Public Expenditure and Reform. This approximately amounts to an average of 50% of a local authority's total capital budget. If local authorities intend to undertake environmental projects, such as flood alleviation works, they must submit proposals to the Office of Public Works (OPW) to obtain this funding. The procurement process for resources to undertake works such as the construction of cycling infrastructure is a similar process whereby proposals are submitted to the National Transport Authority for approval and final decision. Figure 1, which is simplified, illustrates the various state and semi-state actors that local authorities must work with to respond to climate change as well as urban issues, namely, water supply and waste removal. If local authorities had the capacity to raise funds of their own accord, this would enable policymakers to kick-start policies and projects that address climate change in shorter time frames, but this is not the current operational framework. Policymakers view lack of access to funding and financing as a challenge to implementing programmes and actions that would respond to climate change (Interviews, 2019). However, access to finance is only one barrier to action; resources in terms of staff and time are essential.

However, there is a perception by some public servants that the policy tools for implementing climate change adaptation measures are not fully in the hands of local authorities, much less in the hands of civil servants at higher strata of government (Dekker 2018). As reported in Dekker (2018), several interviewees provided anecdotal evidence of attempts to implement policies and programmes, such as permeability and improved public realm, only to be faced with challenges from political actors at higher levels of government. Another factor of policymaking in Ireland is

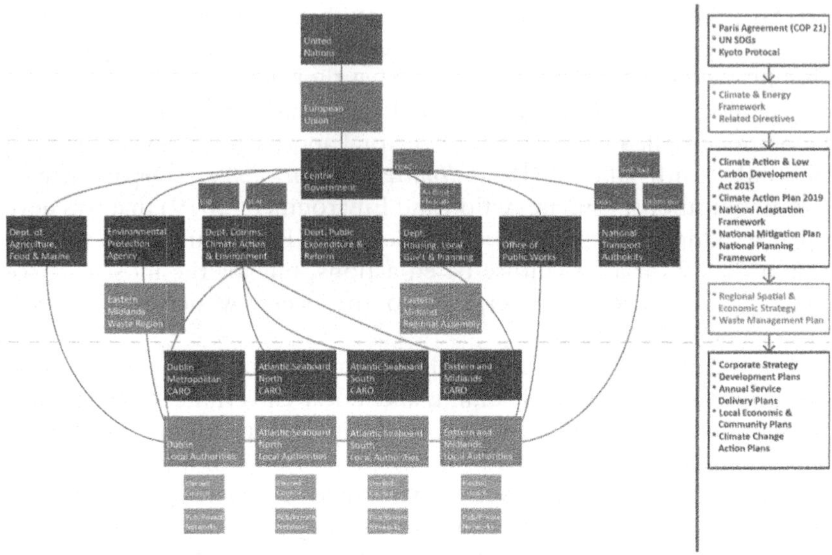


**Fig. 1** Structure of international, national, and local governance as it applies to Ireland

the capacity of individuals to speak directly with ministers in charge of various departments and lobby for change. This access and transparency is useful in that citizens are able to engage with government directly with ministers on issues relevant to them. The impacts of this are perhaps most evident in the Irish Planning System and public consultation.

## Planning Context

The Planning and Development Act 2000 and subsequent amendments regulate planning by local authorities in Ireland. Under the Act, local authorities are required to produce development plans. There have been several planning documents to which local government development plans are connected. The first was the 2002 National Spatial Strategy (NSS) (Department of Environment and Local Government 2002). This was superseded by the 2018 National Planning Framework (NPF) (Department of Housing Planning and Local Government 2018), which sets out the vision for Ireland's growth to 2040. The NSS was intended to guide growth in Ireland. However, the NSS, which served as the reference document for all types of planning, was too detailed in its objectives. This made it cumbersome and an impediment to planning in that it hindered development.

The NPF, Project Ireland 2040, is still detailed and endeavours to redistribute growth from the capital to other parts of the country. There is nominal hope in responding to climate change, as the focus in the NPF in response to climate change is on solutions from the perspective of engineering and economics. This is of concern in that the NPF, as was the case in the NSS, shapes Regional Planning Guidelines, which were revised in 2018 to encapsulate Regional Spatial and Economic Strategies that set out strategies for regional growth. The Regional Spatial and Economic Strategies must in turn be considered in local authority development plans. Additionally, embedded within the development plans are local area plans and local economic and community plans that all local authorities must produce.

Acknowledging that development plans are situated in the context of both national and regional plans, it is evident that formulating a development plan is a challenging task. While local authorities can produce a comprehensive development plan, that plan is subject to national legislation determined by the Department of Housing, Planning and Local Government (DHPLG). As such, endeavours to respond to climate change made by local authorities within development plans are dependent on higher levels of government, recognising the importance of responding.

Another feature of the Irish Planning system is An Bord Pleanála, an appeal body that allows for third parties to lodge objections to planning permissions. This can prolong the process for implementing key projects. Yet, public consultation remains a core requirement for all plans as part of their development process, including those that respond to climate change. Nonetheless, this is viewed as an essential piece of public consultation in the planning process as it enables all citizens to contribute to the system. However, this does not always translate into action.

## Public Consultation

Public consultation is an important feature of planning and policymaking in Ireland. However, as demonstrated by history, perhaps it is in need of reform. Public consultation creates the perception of transparency, but transparency and accessibility are not the same. As such, public consultation is designed for those who know about it, not necessarily for those who need to be consulted and heard. How local authorities formally engage with the public and overcome the challenges, specifically around lack of engagement, calls for a more holistic understanding of the process for public consultation in Ireland.

The public consultation process for the formulation of a development plan is set out in the Planning and Development Act of 2000 (Government of Ireland 2000). These guidelines set the procedures for mandatory public consultation and the corresponding timeline in developing the formal policy. This presents several challenges for planners and policymakers. As has been mentioned already, the guidelines are the same for all local councils in Ireland, irrespective of population size and demographics, geography, and social and economic factors.

All processes related to planning in Ireland, such as public consultation, are intended to be transparent to ensure participation by all is possible. However, transparency does not necessarily translate into participation, as there are challenges with regard to accessibility to information (notably, not everyone has access to the internet and consequently, not everyone is able to access and participate in public consultation processes). In other words, whilst Ireland may have transparent governance, it is not accessible to everyone but is primarily accessible to those who more fully understand the system. This raises the question of how local authorities are able to increase participation and engagement if the information, as transparent as it is, remains inaccessible to the citizens from whom local authorities need to receive input. Further, policymakers face the challenge of stimulating public interest in planning, when attention spans are short and media has the capacity to polarise and often sway public opinion in particular directions.

As discussed, the ability of the citizen to speak directly to ministers and politicians is a trait of the Irish political system and can also impact upon the planning process through elected councillors. Elected councillors represent the voice of citizens and simultaneously are conduits of information. They are tasked with explaining policy decisions to citizens and at the same time conveying citizen expectations to public servants. This proximity to citizens enables elected members to respond immediately. However, remaining a member of a local council is dependent on the ability of councillors to secure votes into the future. Consequently, their ability to influence policy is dependent on their capacity to balance informing constituents of what is needed with, at the same time addressing what their constituents want, thereby securing votes. The competition for votes creates challenges for councillors. While there is an understanding of what is 'good' for the local area, vying for votes may lead to councillors sacrificing their vision for the local authority area in order to retain their seats. This is an area that is exacerbated and certainly not exempt from the centralisation of power across the system.

Policymakers interviewed as part of research that informed the development to the Dublin local authorities climate action plans alluded to central government having a tendency to favour other stakeholders such as developers by creating more onerous procedures for councils to follow in order to develop a project, such as Part 8 which is the process by which local authorities must apply to undertake projects (Government of Ireland 2000). However, as local governments are increasingly recognised to be the level of government best suited to responding to climate change (Allen et al. 2018), understanding how local government operates is increasingly critical to understanding their effectiveness.

## Climate Action Lessons: From Global Cities to Ireland

In the dialogue and debates around how to respond to climate change, cities are increasingly viewed as being the level of government best able to respond and to form policy that will build climate resilience (Betsill and Bulkeley 2007; Dodman 2009; Rosenzweig et al. 2010; Carter et al. 2015). Cities acknowledge their unique position and capacity to engage with people and lead in the process of adapting to climate change and to address the anthropogenic causes of climate change. However, whilst cities may increasingly present themselves as leaders in responding to climate change, there are possible limitations to their capacity to respond effectively, stemming from institutional structures, legislation, and financial capacity, among others. How cities and the local governments that run them respond to these barriers and challenges to achieve climate resilience is valuable in building capacity of cities globally. This section begins by discussing examples of responses being undertaken by cities in other jurisdictions to serve as a benchmark for best practice. Also highlighted is the complexity of the challenge, while showing the opportunities that exist with responding to climate change to achieve co-benefits. Subsequent discussion of the Irish context for local authority climate action is brought into focus and concluding with a case study of Dublin.

Governance of cities is not the same around the world. Local governments are dependent upon regulations and legislation determined by regional and national governments, which influences the financial capacity of cities (Betsill and Bulkeley 2007; Rosenzweig et al. 2010; Dodman 2011; Friedmann 2005). This, however, has not limited local government action on climate change. Given the predominant focus on greenhouse gas (GHG) emissions reduction in the last ten years (Younger et al. 2008), it is not surprising that cities have targeted policies focused on transportation to reduce emissions. Under the umbrella of transportation, cities are focused on improving access to transportation, public transportation routes and services, and providing a variety of transportation options such as trains, light rail systems, and buses. Cities have also promoted alternatives to motorised transport, namely, cycling and walking (active travel). While transportation has been shown to be the largest contributor to GHG emissions in urban areas globally, reducing the number of cars on the road is not the only means by which cities can reduce GHG emissions, improve air quality, and reduce pollution (Younger et al. 2008; Barton 2009; IPCC 2014).

Along with the focus on transportation, cities have focused on the built environment (Tzoulas et al. 2007; Barton 2009; Corburn 2009). Acknowledging that the built environment contributes to emissions through energy use and materials, environmental design has become an emphasis of development. For example, green building guidelines such as Leadership in Energy and Environmental Design (LEED), focused on increasing energy efficiency, using sustainable building materials, and a rating system, are increasingly used by policymakers to encourage developers to construct environmentally friendly buildings (Tzoulas et al. 2007; Corburn 2009). Whilst a focus on green buildings in the residential, commercial, and industrial sectors is beneficial in the policy efforts made by cities to respond to the impacts of climate change, it is not a panacea. Existing guidelines tend to primarily focus on new builds but not existing buildings, which may not be energy efficient or adapted to climate change. Addressing the energy efficiency of existing buildings presents several challenges from costs and suitability of buildings for upgrades to the availability of temporary housing during upgrades, to name a few.

Another means by which cities are responding to climate change is by joining networks that foster friendly competition. Since 1997, an increasing number of partnerships and alliances have been formed by cities, sub-national governments, and philanthropic and community groups, all designed to respond to climate change. These include Climate Alliance, Cities for Climate Protection and the C40 Cities Climate Leadership Group (Betsill and Bulkeley 2007). While initially composed of North American and European cities, the reach of these movements has expanded to include cities in Asia, Africa, Australia, and South America, demonstrating the increasing importance of cities in responding to climate change (Betsill and Bulkeley 2007; Rosenzweig et al. 2010; Geddes et al. 2012; Castán Broto et al. 2013). These initiatives also demonstrate an important role for city and local governments in responding to climate change, namely, their ability to bridge the divide between national governments and citizens (Betsill and Bulkeley 2007).

## Vulnerable Groups and Climate Change

The complexity of the challenges facing cities in addressing climate change demonstrates the need for policies that are comprehensive and integrated to help build resilience and not be limited to the physical environment of the city. The conditions of the social and physical environments in which people live play a significant role in their health and well-being outcomes (Lee and Moudon 2004; Augusto Galvão et al. 2009; Mikkonen and Raphael 2010; Rydin 2012; Rydin et al. 2012). Local governments are tasked with providing high-quality living environments that include housing and economic and social opportunities for citizens (Rees and Wackernagel 1996; Botkin and Beveridge 1997; Rees 1997; Barton 2009). It is this capacity that has played a role in cities being viewed as the appropriate level of government that should address climate change, since they are able to address the day-to-day activities of publics that contribute to climate change (Betsill and Bulkeley 2007; Rosenzweig et al. 2010; Romero Lankao and Qin 2011). More importantly, it is at this level where it is possible to maximise co-benefits that arise from applying a systems approach to climate action.

Addressing poverty is a challenge that is faced by cities globally and requires consideration in responses to climate change. The causes and mechanisms of poverty interact with each other and feed into the 'cycle of poverty' that traps individuals. However, it is primarily constituted as an issue of access to stable employment, affordable and safe housing, education, food and water, and healthcare (Ansari et al. 2003; Barton 2009;

Rydin et al. 2012). The level of access to these basic needs depends of course on where one lives. For example, the urban poor in Mumbai face different challenges from the urban poor of New York. However, in both cities, those with precarious incomes face the risk of falling into a cycle of poverty and vulnerability (Ansari et al. 2003; Rydin 2012; Castán Broto et al. 2013). This cycle is further perpetuated when climate change impacts are considered, especially in cities. Climate change is exacerbating these issues and while policy is actively addressing the physical impacts of climate on the city, policies need to also achieve more depth and a targeted approach (Campbell and Jovchelovitch 2000; Augusto Galvão et al. 2009).

Low and precarious incomes and poverty hinder access to opportunities to improve quality of life via purchasing power, forcing the prioritisation of income expenditure, in other words dictating how household incomes are spent (Lee and Moudon 2004; Augusto Galvão et al. 2009; Mikkonen and Raphael 2010). This in turn compounds the ability of families and individuals to access basic needs, such as housing (Ansari et al. 2003; Lee and Moudon 2004; Augusto Galvão et al. 2009; Mikkonen and Raphael 2010). Access to stable housing plays a key role in livelihood outcomes by reducing vulnerability to numerous risks associated with living outside or precarious housing (Ansari et al. 2003). The solution to this problem is often high-density affordable social housing. Ultimately, climate change perpetuates the vulnerability of an already vulnerable population. Thus, the development of policy needs to give primary consideration to the impact it has on vulnerable groups and mitigate against such increased risks.

## The Irish Context

The Irish Government's *National Mitigation Plan* (Department of Communications, Climate Action and Environment 2017) and *National Adaptation Framework* (NAF) (Department of Communications, Climate Action and Environment 2018b), which were developed under the requirements of the *Climate Action and Low Carbon Development Act 2015*, have tasked key government departments with the development of statutory sectoral adaptation plans. They also charged local authorities with responsibility for developing adaptation strategies that will inform development planning and other statutory plans and with engaging communities at all levels.

Under the NAF, local authorities are viewed as central to implementation of adaptation actions. For example, they are essential in preparing flood emergency response plans. To build the capacity of local authorities to respond to climate change, Local Authority Adaptation Guidelines were produced by DCCAE and Climate Ireland (Gray 2016; Department of Communications, Climate Action and Environment 2018a). These guidelines set out five steps for local authorities to follow as they develop their adaptation plans:

1. Preparing the Ground
2. Assessing the Adaptation Baseline
3. Identifying Future Climate Impacts, Vulnerabilities, and Risks
4. Identifying, Assessing, and Prioritising Adaptation Actions
5. Drafting, Implementing, and Monitoring the Strategy

While the adherence to guidelines is seen in the local authority adaptation plans, some of local authority plans suggest that legislation and guidelines informing the development of the plans may not adequately address local authorities' unique characteristics. This also highlights a dearth of resources both human and financial available in local authorities.

The author of this chapter was responsible for the development of the climate action plans for the Dublin Local Authorities as part of a team of three from Codema, Dublin's Energy Agency. The process was started before the guidelines for local authorities were produced by DCCAE. The plans were developed with consideration of how the Dublin Local Authorities can demonstrate leadership in climate action through their operations and service delivery. The objective of climate leadership and the choice of themes—energy and buildings, transport, flood resilience, nature-based solutions, and resource management—were aimed to facilitate collaboration across departments within the local authorities and in turn to foster ownership of the plans. This was achieved and supported by undertaking workshops with key staff from the relevant local authority departments, as well as carrying out one-to-one meetings. In our review of the existing local authority adaptation plans, it was evident that local authorities were following the local authority adaptation guidelines provided by DCCAE. Interestingly, some local authorities appeared to have utilised aspects of the Dublin plans, which was evident with the use of a similar table layout to present actions highlighting the departments within the local authority responsible for certain actions. The use of thematic action areas was also similar. This may appear to be a non-issue, but when considering the resources in terms of time and staff resources required and available, the need for changes in the process for developing local authority climate action plans becomes evident. To understand the changes needed, it is necessary to consider the process used for the Dublin Local Authorities.

## Development of the Dublin Region's Climate Change Action Plans

Recognising the opportunity to communicate the role of local authorities in Ireland, a clear goal of the Dublin Local Authorities' (DLAs) climate action plans has been to effectively communicate the remit of the local authorities in Ireland. Furthermore, the plans aimed to highlight the desire of the DLAs to do more and to secure a lead in addressing climate action. Responding to climate change is a key priority of the Dublin region's local authorities, as flooding, rising seas, extreme weather, and rising temperatures impact on the region's economic and social growth. Developed in response to international, European, and national calls to action, the Climate Change Action Plans set out the actions that Dublin local authorities will be taking to adapt to and mitigate climate change within the legislative remits of local government in Ireland.

Recognising the capacity of local governments to take a lead on climate action, the DLAs have signed the voluntary European Union (EU) Covenant of Mayors for Climate and Energy (https://www.covenantofmayors.eu/), which commits them to reducing their carbon emissions by 40% by 2030 from 2006 levels and to building resilience to climate change through integrated and collaborative responses.

With this in mind, the DLAs embarked on a process to develop their Climate Change Action Plans. Following the ICLEI Five Milestone process (ICLEI n.d.) and using the *EPA Local Authority Adaptation Strategy Development Guidelines* (Gray 2016), the DLAs, in association with Codema, assessed the risks posed by key climate change impacts and the $CO_2$ emissions from their operations to determine the adaptation and mitigation actions to be undertaken in 2030 (Gray 2016).

Under their Climate Change Action Plans, by 2030, the DLAs will have implemented actions to adapt the region to climate change, such that the risks of flooding, sea-level rise, extreme weather, and rising temperatures are reduced. Furthermore, their Climate Change Action Plans set out a goal that by 2030, $CO_2$ emissions from the DLAs' operations will go further than the 40% reduction target set by the Covenant of Mayors, which would result in them exceeding their emissions reduction targets.

Responding to climate change is an opportunity to strengthen the Dublin region's economic and social competitiveness, as well as its environmental record. Key to this is undertaking a holistic and integrated approach to building climate resilience. Acknowledging this, the Plans were developed by bringing together DLA staff to identify the actions in which they are engaged and the actions they would like to undertake in the future subject to securing adequate financing.

The process for developing the DLA Climate Change Action Plans commenced in April 2017, with the final plans published in September 2019. Codema committed additional time prior to April 2017 to gain approval for undertaking the process. The process was led by an agency outside of the local authorities with one person committed full time to engaging with approximately 100 local authority staff members and helping to write the plans. A number of additional staff were also involved on a part-time basis. Dublin's primarily urban land use (with the exception of agricultural land in Fingal and South Dublin) makes it relatively more straightforward from a policy perspective. Considering the nature of the local authorities across the country, which are more diverse in terms of land uses than the Dublin Metropolitan Region, it becomes evident that significant resources need to be committed to coordinating action. The manner by which coordination of actions was proposed to be achieved was through the establishment of regional offices to support local authorities in their endeavours.

The Climate Action Regional Offices (CAROs) were established in 2018 to support local authorities in the development and implementation of their climate action plans. Additionally, each local authority was to be provided with a team dedicated to climate change. The CAROs were established based on regional climate change risks as identified by the Local Government Management Agency (LGMA) and Cities and County Management Association (CCMA). The proposed groupings were then approved by DCCAE. The resulting four offices are as follows: Atlantic Seaboard North (coastal flooding), Atlantic Seaboard South (sea-level rise), Dublin Metropolitan (urban heat island), and Eastern and Midlands (fluvial flooding). At their establishment, the CAROs were provided with

€10 million in funding over five years. This amounts to €500,000 per year for each CARO to cover staff and administrative costs. Notably, the CAROs have not been endowed with statutory authority, but merely are situated between national and local governments, serving as a conduit (Fig. 1).

As the Dublin plans were developed before the CAROs were created, there are lessons that can be applied to local authorities and sectoral adaptation processes and governance. Critically, the four Dublin Local Authorities were brought together in workshops to discuss their climate change risks and their existing and proposed solutions. These workshops were conducted to facilitate discussion across and within the four local authorities. The purpose of this was to demonstrate that creating climate action plans did not create additional work, but rather provided an opportunity to collaborate with co-workers and to find ways of making available resources (finance, time, and labour) more efficient. Further, when meeting with local authority staff individually, staff were asked about their work, their role within the local authority, and how they wished to make the city a better place to live. This information was used to form groups, identify opportunities for collaboration, and foster ownership of the plans.

## Key Lessons: Collaboration, Integration, and Communication

Workshops are just one tool that enables policymakers to come together and discuss issues and potential solutions. A review of existing local authority adaptation plans suggests that the format of the workshops and the ongoing work outside of the workshops are important. The latter ensures that the outputs of the workshops are used constructively. Bringing together staff from the four DLAs provided an opportunity for staff to talk to counterparts within and across the local authorities (Dekker 2018). Starting a dialogue about what each person and department is doing and would like to do also ensured that the process involved all staff, not just planners. Finally, the workshops served to highlight barriers and challenges that will affect implementation of the plan.

As such, the workshops need to be designed to facilitate discussions that identify problems and support active ongoing problem solving. This will involve bringing stakeholders together to actively solve problems. Critically, the problem may not be directly in their field of expertise but may nonetheless affect them indirectly. By creating this environment, stakeholders will be introduced to individuals that they can contact in their ongoing work. Responsibility for facilitating these workshops could potentially be allocated to the CAROs and DCCAE, because the workshops are a tool for creating an intra-governmental network for climate action. Moreover, all sectoral adaptation plans highlight the key role of local authorities in implementing actions. Further, this network is also essential for communication, both internally and externally to civil society.

Effective communication of climate risks facing Ireland requires the involvement of various central government departments and local authorities, as well as good data to monitor progress on actions. A tool that has not been actively utilised is mapping. Geological Survey Ireland (GSI), located within DCCAE, is tasked with providing open data and maps with information on the geological characteristics of Ireland and interpreting how these interact with proposed projects to build infrastructure and deal with risks such as climate change. The information held by GSI is critical to responding to climate change in a manner that acknowledges Ireland's varying landscape. Mapping may also be used to consolidate where actions are happening and why. This may assist with efforts to achieve collaboration as well as to communicate why actions are being undertaken and how decisions were made. An example to follow can be found in Canada with the Prairie Institute for Climate Change, which has produced an atlas of climate change for the country. This includes climate risk and actions being implemented across the country to respond. Dissemination is an important dimension of this work.

Climate Ireland, a web-based climate information platform developed as a partnership between academia and government, hosts a range of information on climate change in Ireland. The Climate Ireland website could also host a map of the climate actions being undertaken across Ireland and by whom. Further, while the Climate Ireland website has been developed within an academic context with input from policymakers, it retains academic in tone, making it less accessible for public use. If it is to evolve as a tool for public communication, it will need to be made more streamlined, even simplified. This is perhaps an opportunity for collaboration between the CAROs and the National Dialogue for Climate Action to develop a platform that effectively communicates what is being done in Ireland by local authorities and government departments to address climate change.

## Conclusions

With local governments increasingly recognised to be the level of government best suited to responding to climate change (Allen et al. 2018), an understanding of how local government operates is critical in gauging their overall effectiveness. This is possibly one of the key lessons to be learned from local authorities in Ireland. They have taken often difficult action within constraints and have emerged as leaders, especially in the field of climate change. The ability of local authorities to deliver on the national government's expectation that they will support communities in building climate resilience faces barriers. Local authorities in Ireland, much like their counterparts globally, can take a leadership role in climate action. Already, local authorities in Ireland have demonstrated their capacity for leadership through the implementation of programmes to collect data relevant to climate change through smart cities initiatives. These initiatives have involved collaboration with private sector and academia. Limited remits and financial capacity can be addressed by processes that foster collaborative action. The CAROs are a resource for local authorities to facilitate collaboration within and across the local authorities in each CARO region. Whilst lack of resources in terms of staff and finances are a challenge for the CAROs, as they are for local authorities, the networks that they provide can be used to highlight the commitment of local authorities to respond to climate change. However, in order for the responses to achieve the necessary transition, more funding is required from central government.

## Bibliography

Allen, M. et al. 2018. *IPCC Global Warming of 1.5 oC: Summary for Policymakers SPM, IPCC.*

Ansari, Z., et al. 2003. A Public Health Model of the Social Determinants of Health. *Sozial- und Präventivmedizin* 48 (4): 242–251. https://doi.org/10.1007/s00038-003-2052-4.

Augusto Galvão, L.C. et al. 2009. Panel Climate Change and Social Determinants of Health: Two Interlinked Agendas 1 Pan American Health Organization-Area of Sustainable Development and Environment. https://doi.org/10.1177/1757975909103761.

Barton, H. 2009. Land Use Planning and Health and Well-Being. *Land Use Policy* 26: 115–123. https://doi.org/10.1016/j.landusepol.2009.09.008.

Betsill, M., and H. Bulkeley. 2007. Looking Back and Thinking Ahead: A Decade of Cities and Climate Change Research. *Local Environment* 12 (5): 447–456. https://doi.org/10.1080/13549830701659683.

Botkin, D.B., and C.E. Beveridge. 1997. Cities as Environments. *Urban Ecosystem* 1: 3–19.

Campbell, C., and S. Jovchelovitch. 2000. Health, Community and Development: Towards a Social Psychology of Participation. *Journal of Community & Applied Social Psychology* 10 (4): 255–270. https://doi.org/10.1002/1099-1298(200007/08)10:4<255::AID-CASP582>3.0.CO;2-M.

Carter, J.G., et al. 2015. Climate Change and the City: Building Capacity for Urban Adaptation. *Progress in Planning* 95: 1–66. https://doi.org/10.1016/j.progress.2013.08.001.

Castán Broto, V., B. Oballa, and P. Junior. 2013. Governing Climate Change for a Just City: Challenges and Lessons from Maputo, Mozambique. *Local Environment* 18 (6): 678–704. https://doi.org/10.1080/13549839.2013.801573.

Corburn, J. 2009. Cities, Climate Change and Urban Heat Island Mitigation: Localising Global Environmental Science. *Urban Studies* 46 (2): 413–427. https://doi.org/10.1177/0042098008099361.

Dekker, S. 2018. *Cities Leading Climate Action: Urban Policy and Planning*. London: Routledge.

Department of Communications, Climate Action and Environment, D. 2017. National Mitigation Plan (July), 1–198. Available at: http://www.dccae.gov.ie/documents/National.Mitigation.Plan.2017.pdf

———. 2018a. *Local Authority Adaptation Strategy Development Guidelines*. Dublin.

———. 2018b. National Adaptation Framework Planning for a Climate Resilient Ireland, 1–110. Available at: https://www.dccae.gov.ie/documents/National.Adaptation.Framework.pdf

———. 2019. *To Tackle Climate Breakdown*. Dublin.

Department of Environment and Local Government, D. 2002. *National Spatial Strategy for Ireland 2002–2020 People, Places and Potential*. Dublin.

Department of Environment Community and Local Government, D. 2015. *Climate Change and Low Carbon Development Act 2015*. Available at: http://www.oireachtas.ie/viewdoc.asp?fn=/documents/bills28/bills/2015/215/b215d.pdf

Department of Housing Planning and Local Government, D. 2018. *National Planning Framework*. Dublin.

Dodman, D. 2009. Blaming Cities for Climate Change? An Analysis of Urban Greenhouse Gas Emissions Inventories. *Environment and Urbanization* 21 (1): 185–201. https://doi.org/10.1177/0956247809103016.

———. 2011. Forces Driving Urban Greenhouse Gas Emissions. *Current Opinion in Environmental Sustainability* 3 (3): 121–125. https://doi.org/10.1016/j.cosust.2010.12.013.

Friedmann, J. 2005. Globalization and the Emerging Culture of Planning. *Progress in Planning* 64: 183–234. https://doi.org/10.1016/j.progress.2005.05.001.

Geddes, A., et al. 2012. Migration, Environmental Change, and the "Challenges of Governance". *Environment and Planning. C, Government & Policy* 30: 951–967. https://doi.org/10.1068/c3006ed.

Government of Ireland. 2000. *Planning and Development Act 2000.* Available at: http://www.irishstatutebook.ie/eli/2000/act/30/enacted/en/pdf. Accessed 24 Feb 2020.

Gray, S. 2016. *Local Authority Adaptation Strategy Development Guideline.* Available at: www.epa.ie. Accessed 24 Feb 2020.

Houses of the Oireachtas. 2019. *Report of the Joint Committee on Climate Action. Climate Change: A Cross-Party Consensus for Action.* Dublin: Houses of the Oireachtas.

ICLEI. n.d. *Five Milestones of Emissions Management – ICLEI USA.* Available at: https://icleiusa.org/programs/emissions-management/5-milestones/. Accessed 24 Feb 2020.

IPCC. 2014. *Summary for Policymakers—IPCC Fifth Assessment Report.* Available at: https://www.ipcc.ch/report/ar5/wg3/summary-for-policy-makers/. Accessed 24 Feb 2020.

Lee, C., and A.V. Moudon. 2004. Physical Activity and Environment Research in the Health Field: Implications for Urban and Transportation Planning Practice and Research. *Journal of Planning Literature* 19 (2): 147. https://doi.org/10.1177/0885412204267680.

Mikkonen, J., and Raphael, D. 2010. *Social Determinants of Health: The Canadian Facts.* Available at: http://listserv.paho.org/Archives/equidad.html. Accessed 24 Feb 2020.

Rees, W. 1997. Urban Ecosystems: The Human Dimension. *Urban Ecosystem* 1 (1): 63–75. https://doi.org/10.1023/A:1014380105620.

Rees, W., and M. Wackernagel. 1996. Urban Ecological Footprints: Why Cities Cannot Be Sustainable – and Why They are a Key to Sustainability. *Environmental Impact Assessment Review* 16 (4–6): 223–248. https://doi.org/10.1016/S0195-9255(96)00022-4.

Romero Lankao, P., and H. Qin. 2011. Conceptualizing Urban Vulnerability to Global Climate and Environmental Change. *Current Opinion in Environmental Sustainability* 3 (3): 142–149. https://doi.org/10.1016/j.cosust.2010.12.016.

Rosenzweig, C., et al. 2010. Cities Lead the Way in Climate-Change Action. *Nature* 467: 909–911. https://doi.org/10.1038/467909a.

Rydin, Y. 2012. Healthy Cities and Planning. *The Town Planning Review* 83 (4): xiii–xviii. https://doi.org/10.2307/41509860.

Rydin, Y., et al. 2012. Shaping Cities for Health: Complexity and the Planning of Urban Environments in the 21st Century. *The Lancet* 379 (9831): 2079–2108. https://doi.org/10.1016/S0140-6736(12)60435-8.

Tzoulas, K., et al. 2007. Promoting Ecosystem and Human Health in Urban Areas Using Green Infrastructure: A Literature Review. *Landscape and Urban Planning* 81 (3): 167–178. https://doi.org/10.1016/j.landurbplan.2007.02.001.

Younger, M., et al. 2008. The Built Environment, Climate Change, and Health. Opportunities for Co-benefits. *American Journal of Preventive Medicine* 35 (5): 517–526. https://doi.org/10.1016/j.amepre.2008.08.017.

SECTION III

# Media, Communication and Society

# Ecological Modernisation, Irish-Style: Explaining Ireland's Slow Transition to Low-Carbon Society

*Declan Fahy*

## Introduction

Ireland's response to climate change has appeared confused and contradictory. In several respects, the country's recent environmental policies and initiatives have been pioneering and progressive. Ireland became the first country in the world to divest from fossil fuels and the second country in the world to declare a climate emergency. The country has implemented a carbon tax, enacted a framework of legislation to address climate change, and vowed to protect poor and vulnerable citizens as the country aims to transition to a society with lower carbon emissions. It has banned onshore fracking and has promised to increase the amount of energy derived from renewable energy sources. It sought diverse public viewpoints on climate change policy as part of a Citizens' Assembly on climate change, one of

D. Fahy (✉)
School of Communications, Dublin City University, Dublin, Ireland
e-mail: Declan.Fahy@dcu.ie



D. Robbins et al. (eds.), *Ireland and the Climate Crisis*, Palgrave Studies in Media and Environmental Communication,
https://doi.org/10.1007/978-3-030-47587-1_8

the first examples in the world of this form of climate-focused public deliberation. In 2019, the Government published the country's most ambitious plan to decarbonise the economy to respond to what it called a 'climate breakdown' (Government of Ireland 2019; Little and Torney 2017).

But in several other respects, Ireland's policies and performance on environmental issues have been regressive and unsustainable. It has pursued policies to grow the country's cattle herd, even though agriculture is the sector that accounts for the largest amount of greenhouse gas emissions—almost one-third of the country's total and three times greater than the European Union (EU) average. It has been slow to phase out its single coal-burning and three peat-burning power stations. It has been slow to implement, or has failed to implement, EU environmental law. Fundamentally, for a developed country, its record on sustainability has remained, for decades, by international standards, across a range of environmental science measures, abysmal (Burck et al. 2020; Torney and O'Gorman 2019; Government of Ireland 2019; Little and Torney 2017; Little 2017). It is easy to conclude, therefore, that the country is, in the colloquial Irish phrase, speaking out of both sides of its mouth, promising environmental reforms while prioritising the economic development that caused enormous ecological damage.

However, this chapter argues that these confused and contradictory policies and processes make more sense when they are understood within the context of one distinct model of sustainable development, one called ecological modernisation, which at its core sees economic growth as compatible with ecological protection. Scholars have noted that Ireland's overall pattern of development has conformed to ecomodernist ideas (Tovey 1992; Leonard 2007; Pepper 1999). This chapter presents a historicist argument that the process of ecological modernist development in Ireland has been slow and ineffective, as policymakers have for decades consistently prioritised economic development over ecological conservation. Ireland's plodding path to sustainable development, this chapter argues, has been the result of a set of historical forces that have shaped and moulded the country's relationship with the natural world, with agriculture an especially powerful force in the country's economic and social history. As this chapter explains, while Ireland has tried to pursue a path of sustainable development, fundamental factors in Irish history have made it difficult for ecological modernist ideas to take root. These factors have meant that a weak version of ecological modernisation has prevailed in

Ireland, a feeble collective response to, and preparation for, a climate-challenged society.

## Ecological Modernisation and Social Change

Ecological modernisation, or ecomodernism, or eco-restructuring, is a theory of social change that describes the process of sustainable development. It argues that economic development can proceed in tandem with ecological protection. More fundamentally, it argues that economic growth depends on the preservation of the environment, the material base of society, and the provider of natural resources on which human society depends. Opposed to ideas of degrowth or demodernisation, which argue that the way to address environmental problems is to slow down or halt economic growth, ecomodernism rejects apocalyptic visions of the future and sees environmental problems as challenges that can catalyse systemic social reform. The set of ideas that constitute ecomodernist thought was developed by European sociologists in the 1980s, and the ideas were codified in the 1987 UN report *Our Common Future*, usually referred to as the Brundtland Report after its chair, former Norwegian Prime Minister Gro Harlem Brundtland. It remains a landmark document in the history of sustainable development, which it defined (UN 1987: 41) as 'development that meets the needs of the present without compromising the ability of future generations to meet their own needs'. Written before the full realisation of climate change's consequences, the report, its authors noted (UN 1987: 1), 'is not a prediction of ever increasing environmental decay, poverty, and hardship in an ever more polluted world among ever decreasing resources. We see instead the possibility of a new era of economic growth, one that must be based on policies that sustain and expand the environmental resource base'.

In order for such a sustainable society to develop, ecomodernists argue, political and economic systems must be reformed along ecological lines. That means all the core systems and institutions of modern society—market economies, industrial production, welfare states, agricultural production, and scientific and technological institutions—must be restructured to protect the environment. As a large-scale theory of social change, ecomodernism examines and explains how these systems and institutions of modern societies can be reformed. As a specific political programme, ecomodernism directs policies that conserve the environment. For the environmental politics scholar John Dryzek, ecomodernism aims to reform, not replace, the capitalist economic system. It aims at nothing less, he wrote (2013: 145), than 'the ecological restructuring of capitalism'.[1]

Central to that restructuring is partnership–between governments, companies, moderate environmentalists, and scientists. They are all motivated, argued Dryzek (2013: 174), 'by the common good or the public interest, defined in broad terms to encompass economic efficiency and environmental conservation'. Within this broad partnership, policies are forged in a consensus-based decision-making process, with governments, businesses, scientists, and environmentalists involved in planned policy interventions that affect investment patterns, planning decisions, and research funding. In order for such a corporatist partnership between government, business, and civil society social partnership to work, the partners must coordinate around clear policy interventions. Governments should factor environmental considerations into all public policies, set strong industrial regulations, and provide companies with incentives to innovate. Governments should help catalyse new technologies that reduce resource consumption and emission production, as well as providing consumers with supports to adopt more environmentally friendly behaviour. Not least, public support is vital to the ecomodernist project (Weale 1992; Mol and Spaargaren 2000; Dryzek 2013).

These ecomodernist ideas have been criticised for decades. Critics argue that ecomodernism is too optimistic about new technologies, too reformist in the face of urgent problems, too focused on the effects, rather than the causes, of unsustainable development, too uncritical of capitalism with its demand for orthodox economic growth that they argue is inherently incompatible with environmental conservation. In response, two major ecomodernist thinkers, Arthur Mol and Gert Spaargaren (2000), argued that capitalism was constantly changing in the face of different circumstances—and environmental concern has been one trigger for change. They also argued that production and consumption practices have changed to be more sustainable. In their view, other alternatives to the current economic order have proved unfeasible. The green political economist John Barry (2007: 447) argued that any conception of 'an alternative economy and society must be based in the reality that most people (in the West) will not democratically vote for a completely different type of society and economy'. In addition, Barry argued that ecomodernism had value as a foundation for the future development of more radical environmental policies. Ecomodernism, he argued, allowed sustainability to be integrated into the core political imperative of economic growth. Barry

(2007: 460) also wrote that viewing ecomodernism as a starting point for a model of green political economy had the benefit of removing ideas of degrowth, which he viewed as holding back 'the theoretical development of a positive, attractive, modern conceptualisation of green political economy and radical conceptualisations of sustainable development'.

In his influential *The New Politics of Pollution* (1992), political scientist Albert Weale argued that, since the 1980s, Germany and the Netherlands have been examples of countries that have made significant attempts to transform their countries along ecomodernist lines. There were particular characteristics of these countries, he argued, that meant ecomodernist ideas resonated with policymakers and the public. They had interventionist states that supported policy innovation around the environment. They provided major public investment and subsidies to allow industries adopt more sustainable production processes. They integrated environmental policies across government departments. They were corporatist, in that elites in society and the state and civil society together set the framework for public policy. Their citizens were disposed to environmental protection in their attitudes and were keen to participate in social transition, a public disposition formed in part by the consciousness-building efforts of green politicians, the environmental movement, and environmental journalism.

Yet despite these advances in particular countries, despite the optimistic promises of ecomodernism, no country has yet transformed itself into a fully realised example of an ecomodernist state. Some states, however, have had more thought-out and effective versions of ecomodernism than others. In an insightful paper, Christoff (1996) argued that there are two broad ways of conceptualising ecomodernism. Weak ecomodernism—or what he was tempted to call false ecomodernism—has had as core characteristics the prioritisation of economic interests, a focus on technological solutions, and a closed and corporatist style of decision-making with limited public participation. Strong ecomodernism, by contrast, has had as core characteristics a prioritisation of ecological values, a focus on broad systemic changes, and a commitment to communication to involve citizens in deliberative decision-making. Christoff argued that weak and strong ecomodernism could be understood as poles on a continuum, with some states further along the continuum towards strong ecomodernism than others. There have been multiple factors that determine where a country falls in this continuum. According to Cohen (1998), decisive factors include the economic organisation of a society and what he called a country's cultural capacity for ecomodernism. He argued that countries more likely to undertake strong ecomodernism have strong public commitment to science and a strong environmental consciousness.

Ireland's policy on sustainable development is built on ecomodernist ideas. The sociologist of the environment Hilary Tovey (1992) argued that ecomodernism was what she called Ireland's official environmentalism, a perspective led by economists, environmental scientists, agricultural scientists, established environmental organisations, and public sector workers involved in environmental-related organisations. Yet the country's atrocious record of environmental protection has demonstrated that ecomodernism in Ireland has been a clear manifestation of weak ecomodernism, or false ecomodernism, in which policymakers and businesses have prioritised short-term economic aims at the expense of ecological concerns. In a strong critique, Leonard argued (2007: 469) that Ireland has produced 'an illusion of ecomodern progress', a state of affairs that resulted from the neo-liberalist economic policies of successive modern Irish governments. However, Ireland's weak ecomodernism can be explained with reference to other historical factors and forces, which have interacted to create the conditions in which ecomodernism as a theory of social change has not mapped well onto the particular history of Ireland.

## Irish Agriculture and the Iron Law of Climate Policy

The first factor is the relationship between modernisation and Ireland. Modernisation is an influential large-scale theory of development that argues that all countries, though moving at different rates, are in the process of becoming modern industrial societies (Share et al. 2007). However, as sociologists and historians have argued, modernisation theory, based largely on the histories of Britain and some continental European countries, has been unable to give a full picture of Irish social change. Ireland was a colonial country for centuries, its economic and social development shaped by Britain. When Ireland gained independence in 1922, it had a very small industrial sector and its early industrial growth was driven by protectionist economic policies that shielded Irish goods from competition from cheaper exports. From the late 1950s, new economic policies meant Ireland industrialised rapidly and introduced export-focused policies to reintegrate itself into global markets (O'Malley 1992). As a result of this industrial history, Ireland has often been called a 'late moderniser'.

For some sociologists, this description was pejorative, as it presented Ireland as somehow 'less than modern'. Instead, a more constructive way of viewing modernisation, wrote Share, Tovey, and Corcoran (2007: 47), is to understand that there are 'different varieties of modernisation'.

Therefore, any explanation of the Irish variety of modernisation, according to Share, Tovey, and Corcoran (2007), must not begin, as in other countries, with a conventional account of industrialisation. It must begin instead with an account of agriculture and agricultural change. The crucial influence of agriculture on Ireland's economic and society history is the second factor that has shaped Ireland's attempts at sustainable development. As a colony of Britain, Ireland specialised in agriculture, especially the production of live beef cattle for export. As a result, Irish food has been a global commodity since the seventeenth century, moving through the British Empire's trade routes and colonial circuits. After independence, agriculture remained central to Ireland, with the main post-colonial industries being the agriculture-based ones of brewing, distilling, and food processing. 'Since the foundation of the State in 1922', wrote political scientist Maura Adshead (1996: 604), 'the agricultural sector in Ireland has been accorded a special status, both for its own strategic importance and as a key industry for the promotion of economic growth'.

As Tovey (1992) perceptively noted, debates about the environment in Ireland were a proxy for debates about the place of agriculture in Irish society. Until the middle of the twentieth century, Ireland was a rural-based society with a focus on smaller, separate family farms. From then, agriculture developed as a commercial industry, focused on providing food for the export-driven food industry, becoming in the process a major sector of the Irish economy (Tovey 1992). As part of this development, Irish agriculture has been shaped by three economic and technological forces. The first was productionism, in which agriculture was narrowly viewed as the production of food to the neglect of other social functions, such as providing jobs or protecting the environment. The second was commercialisation, whereby Irish agriculture became a producer of products to be sold to global food markets as part of the global agribusiness industry that has come to dominate agriculture worldwide. The third was scientisation, a process in which innovations in agricultural science were applied to increase farming production. As a result, wrote Share, Tovey, and Corcoran (2007), Irish 'agriculture can no longer be thought of as a set of independent, separate farming units but instead as part of an extensive, even globalised, food system'. In a stronger left-wing critique,

Shanagher and Brereton (2019: 2) characterised contemporary Irish farming as a 'capitalist cattle-economy'.

Because of agriculture's economic importance, Irish farmers have wielded great political influence. In the middle of the twentieth century, farmers organised into representative organisations, including the Irish Farmers Association and the Irish Creamery Milk Suppliers Association, which have influenced, through their lobbying efforts, the agricultural policies of successive governments. So great was the influence of the agricultural sector, argued Adshead (1996: 83), that farmers and the government came to operate within a political structure she called a 'closed policy community'. In such a community, policymaking occurred in closed-circle discussions between interest groups, ministers, and senior civil servants who shared ideological viewpoints and priorities. The public and other interest groups were essentially excluded. Because this policy community has had such a long history in Ireland, with an older and strategically important ministry at its heart, it has been difficult for other organisations and interests to confront these entrenched practices. The agricultural lobby has remained a strong voice and powerful interest at a national and EU-wide level.

That influence can be seen in Ireland's climate policy negotiations with the European Union. The EU has enacted more and more green-focused policies since the 1980s, but Ireland has managed to resist fully implementing these policies (Torney and O'Gorman 2019). Part of the explanation for this state of affairs is that Ireland, according to Coghlan (2007), has played what political scientists call a 'two-level game' around climate policy. It engaged in one game, at the international policy level, and a second game, at the domestic policy level. When policy was discussed in the EU, Ireland committed to limiting greenhouse gases and discussed the policies that would effect this limitation. But when policy was discussed in Ireland, interest groups such as the farming and business lobbies exerted great influence and climate policies, agreed at the international level, became diluted at the domestic level. Coghlan (2007) concluded that Irish climate policy was weak because it was unable to 'win' at both levels—that is, it was not possible to develop climate policy that would satisfy the international and domestic interests. Domestic agendas dominated and so economic interests won out.

Evidence of Irish policymakers operating at these two levels is found in the public pronouncements of two recent Taoisigh. In 2015, Enda Kenny told a journalist at the UN climate summit in Paris that Irish agriculture should be considered a special case when it comes to emissions reductions. It was exceptional, he argued, because it produced food in a more sustainable way than other countries and because short-term economic growth was needed to create the funds for investment in research to develop long-term sustainable agriculture. Soon after, Kenny gave a speech to the summit in which he stressed Ireland's commitment to climate action (Little 2017). In 2019, Leo Varadkar argued that Irish food production should be considered differently to other emission sources because there is a global demand for food and the country exports 90 percent of the food it produces. But as Bindi and Olesen (2011) outlined, there will likely be two strong trends in European agriculture that will profoundly affect Ireland. The first trend is the continued move towards globalisation of markets for agricultural products. The second is an increased focus on the nature of the goods produced by agriculture, including the social and environmental impacts of those goods and how they are produced. Irish agriculture appears to be locked into these trends, so any reform of Irish agriculture will occur in a policy context where agriculture remains an important part of Ireland's outward-focused economy.

A central part of efforts to address agricultural emissions is the research of Teagasc, the agricultural research agency. It plans to apply scientific solutions to reduce greenhouse gas emissions, such as using novel fertilisers to reduce nitrous oxide, as well as planting forests to sequester carbon dioxide (Lanigan 2017). However, such a narrow focus on technological solutions is a feature of weak ecomodernism. It can be understood also within a wider historical context in which science has not been central to Irish culture. As the science studies scholar Steven Yearley noted, scientific rationalism did not feature as a pillar of the new Irish state from 1922. 'Science', he argued (1995: 178), 'was not championed as an exemplary component of a "positive" or "progressive" culture', and 'technological planning was far from the state's mind even though a little official attention was focused on industry, with state ventures in electricity generation, the production of industrial alcohol, sugar processing, and the exploitation of peat as a fuel source'. The social position of science changed in the closing decades of the twentieth century, however, as there was a surge in political support for science and technology. However, rather than presenting science as a driver of social change or a way to restructure society along environmental lines—the Environmental Protection Agency (EPA) was not established until 1992—the benefits of science were framed narrowly in economic terms. As one long-time writer on Irish science policy and science communication, Brian Trench (2009, p.4) wrote, science, especially the fields of biotechnology, information technology, and biosciences, became a focus for Irish industrialisation, 'based largely on foreign direct investment in export-led high-technology manufacturing and high-knowledge services and concentrated on the bio-pharmaceutical and information technology sectors'. Irish science served economic growth over ecological conservation.

As a result of these historical factors, Ireland has experienced a particular manifestation of the 'iron law of climate policy'. The environmental policy scholar Roger Pielke, Jr., conceptualised this term to refer to the inevitable outcome that occurs when policymakers must choose between economic objectives and environmental protection. In his book *The Climate Fix* (2010: 46) he argued that, 'if there is an iron law of climate policy, it is that when policies focused on economic growth confront policies focused on emissions reductions, it is economic growth that will win out every time'. In Ireland, agriculture is a major greenhouse gas emitter and a major economic motor. When forced to choose, policymakers have prioritised economics. The iron law has held in Ireland.

## Ireland's Weak Ecological Consciousness

As well as the country's economic structure, with the importance of export-driven agriculture, another factor in Ireland's weak ecomodernism has been the country's historically low level of ecological consciousness. In order for ecomodernist ideas to take root, they, like other political ideas, as the communications scholar Brian McNair wrote, need to resonate with the general public in order to have influence. 'As a general rule', he wrote (2011: 29), 'the effects of political communications of whatever kind are determined not by the content of the message alone, or even primarily, but by the historical context in which they appear, and especially the political environment prevailing at any given time'. In the 1980s, for example, there was a surge in public support, aided by Green party advocacy, for the environment in Germany and the Netherlands as the countries moved to implement ecomodernist ideas (Weale 1992). In that decade in Ireland, people ranked in surveys the environment as one of their lowest topics of concern, a trend that was especially pronounced when the issue was presented as a trade-off between environmental and economic goals (Coyle 1994, cited in Share et al. 2007). That low level of interest has continued, with Irish citizens' interest in the environment trending below the European average for several years (Little 2017).

The country's overall low environmental sensibility can be explained by the fact that, in Ireland, three powerful shapers of public opinion—environmental journalism, green politics, and the environmental movement—have had marginal and ineffective presences in public life. In other western countries, environmental journalists have been a major force for shaping public around the environment. Through their regular reporting, these specialists place and keep the environment on the public radar, emphasising through consistent coverage that the environment is an important issue of concern, one that policymakers and voters should think and care about (Fahy 2017a). However, there has been a historical lack of specialist environmental journalists in Irish media, especially compared with the larger amount of reporters focused on politics or business. This lack of specialists has been a reason for the fact that coverage of climate change in Ireland has received less coverage than other European countries. When the environment was covered, reports tended to focus on events, such as the publication of Intergovernmental Panel on Climate Change (IPCC) reports and the convening of international climate summits. Only *The Irish Times* has had a record of consistent coverage, including the publication of editorials that argued for increased climate action (Robbins 2018; Fox and Rau 2018). The view of sustainable development in elite newspaper coverage, one study argued, was ecomodernist in that existing institutions were presented as being able to format a response to the systemic issue of climate change (Wagner and Payne 2017). Yet these ideas, propagated by Irish political and economic elites, the narrow set of sources consulted by journalists, amounted to little more than weak ecomodernism.

The lack of influence of environmental journalism has stood in marked contrast to the influence of journalism more generally on Irish society. As the media historian Mark O'Brien noted, Irish journalism in the twentieth century has been a major force in the country's social and cultural modernisation. He argued in *The Fourth Estate* (2017) that Irish journalism played a vital role in leading progressive social change around social issues, such as divorce, abortion, and marriage equality. It played a role in incorporating feminist ideas into Irish society, as well as being involved in reporting the abuses of the Catholic Church, which had a dominant influence over Irish public opinion for much of the twentieth century (O'Brien 2017). Missing, however, from that movement to modernise has been a form of environmental modernisation akin to social and cultural modernisation. In a notable comparison, the farming community has formed its own communication networks in which it has discussed vital issues. A key part of this network has been the *Irish Farmers Journal*, which, since its foundation in 1948, has worked, as it described its own role, 'with Irish farmers and the agricultural industry, to encourage and sustain a prosperous farm economy in Ireland'. The newspaper is owned by a trust, with the sole mandate to 'provide the best possible technical support, market information and news to the agricultural sector' (Irish Farmers Journal 2019).

The collective voices of Green politicians have been a force that can shape public opinion on the environment (see chapter "The Party Politics of Climate Change in Ireland" by Little, this volume). Elected green politicians can, through their policies and media appearances, put environmental issues on the public agenda and can persuade citizens to see environmental problems and issues in a certain way, leading, over time, to the formation of a critical mass of public opinion that can force policy change. Green politics can also catalyse social movements, to increase citizen participation in politics that can put pressure on governments and industries to change. But the Green Party has been a marginal presence in modern Irish history. It was founded in 1981 as The Ecological Party of Ireland and had its first general election success in 1989 when Roger Garland was elected to the Dáil for South Dublin. It increased its Dáil representation to two seats in 1997, six seats in 2002, and six seats in 2007 when it joined as a minority partner in the Fianna Fáil government, the first time the party had been in government. After the general election in February 2020, out of 160 members of Dáil Éireann, the Greens had 12 TDs. The party support has been concentrated in Dublin and has tended to be greatest among the younger and wealthier parts of the population (Marsh 2020). And when the party was in government, it presented two different, but related, views of how climate change could be addressed. One then Minister, Eamon Ryan, portrayed climate change around the ecomodernist issues of new technologies, energy infrastructure, and renewable energy. The other minister, John Gormley, stressed lifestyle changes to tackle the issue. The result was a party that lacked a clear, unified ministerial philosophy on what to do to address climate change (Robbins 2020).

Environmental movements, internationally, have been a third motor of public influence. But not in the north and south of Ireland, 'Whether measured by membership or influence on policy and politics, it emerges as one of the weakest in Europe' (Doran and Barry 2009: 322). The movement as a whole has consisted of large environmental organisations and smaller conservation groups. The worldview of Irish environmentalists can be categorised as one that is opposed to mainstream ecomodernism, a perspective Tovey (1992: 529) called 'populist environmentalism'—an independent movement of dissent by local groups of people in response to 'the dominant ideologies of modernisation, development and growth'. Environmentalists have mobilised multiple times in recent decades in protest against individual state-backed development projects. This can be seen, as Leonard argued (2007), in the protests against proposals for incinerators in the 1990s, the Shell to Sea campaign (in which local communities in the west of Ireland protested against an onshore gas pipeline), and protests against road development around the historic sites the Glen of the Downs and the Hill of Tara. But the most notable environmental movement in Ireland was the anti-nuclear movement in the 1970s, which mobilised local opposition to the state's plans to build a nuclear power station at Carnsore Point in Wexford, plans that were scrapped in the face of such opposition (Baker 1990, cited in Share et al. 2007). A unifying feature in the various elements of the Irish environmental movement has been a general opposition to ecomodernist development.

## A Strong Ecomodernist Future?

In conclusion, Ireland has attempted to forge a path to sustainable development by following an ecomodernist form of social change. However, Ireland's ecomodernism has prioritised economic gains over ecological values, a pattern that can be explained with references to a series of interconnected factors and forces in its history. Ireland has had a historical reliance on agriculture as a vehicle for economic growth, as well as strong farming and business interest groups that have shaped environmental policy within a closed policy community, a network of insiders that has largely excluded environmental organisations, which themselves have had a marginal presence in policy formation. Ireland has had a low level of public interest in environmental issues, a stance shaped and reflected by the low representation of elected Green politicians and the lack of sustained journalistic attention to environmental issues. As a cumulative result, the theory of ecological modernisation has been, at best, an awkward fit with the features of Irish history. The result has been a weak or false ecomodernism, or what might be called ecomodernism, Irish-style: promises of ecological prioritisation undercut by a historically determined focus on agriculture as a driver of economic growth.

Ireland remains on an ecomodernist path. And there is evidence of some movement from a weak to a strong ecomodernism. This is most evident in Ireland's significant progress on the deployment of renewable energy (Torney and O'Gorman 2019). Also, the Citizens' Assembly demonstrated its value as one method of citizen participation in deliberative decision-making. The Green Party has experienced recent electoral successes, which, along with student involvement in climate protests, are tentative evidence of a greater ecological consciousness (see chapter "The Changing Faces of the Climate Movement in Ireland" by Lorna Gold's, in this volume). The Climate Action Plan in 2019, the most far-reaching document in Irish climate policy history, stressed the need for systemic social change in the face of climate change, emphasising that all the institutions of society need to be restructured along ecological lines, a textbook example of ecomodernist doctrine. That plan stressed that environmental protection underpinned economic development, presenting social restructuring as 'important not just for Ireland's long-term economic and societal interests, but also in relation to the attractiveness of Ireland as a location for Foreign Direct Investment, as a tourism destination, and as a source of safe, high-quality agricultural and food products' (Government of Ireland 2019: 9). The plan's view on agriculture is ecomodernist in that it seeks to protect the economic value of sustainable farming. In language that is a fine example of political equivocation, the plan (2019: 98) described the long-term future of agriculture, which it said faced the formidable challenge of meeting 'the national policy objective of an approach to carbon neutrality which does not compromise our capacity for sustainable food production'. It's language that shows that the historical tension between agriculture and the environment in Irish environmental debates has still not been resolved.

That lack of resolution on farming is emblematic of a larger failure. Missing from the Climate Action Plan, as has been missing from much Irish social thought about the environment, is a clear vision of a future Irish society, beyond the hopeful prospect of a country with less carbon emissions. The climatologist Mike Hulme argued in *Why We Disagree About Climate Change* (2009) that the challenges of climate present societies with a chance to discuss what kind of societies they want to be and how those societies can be created through the political, social, economic, and cultural responses to climate change. Ireland has not yet had such a broad public discussion, except for the weak or feeble ecomodernism communicated by successive political and business elites. As the sociologists of science Harry Collins and Robert Evans (2017) wrote in reference to the place of science in democracies, societies must choose to be modern, a form of collective reasoning they called elective modernism. Their insight has application for Ireland, a country that must now decide to end its calculated ambivalence over ecological values and elect be ecomodern. Ireland, sociologists and historians have long argued, has been a late economic moderniser and a late social moderniser. Ireland still has the potential to be a late, but strong, ecological moderniser.

## Note

1. There is a vast scholarly literature on ecomodernism. Writers have examined ecomodernism from various perspectives and different critical approaches, viewing ecomodernism as, variously, a theory of social change; an ideology; a discourse or way of talking about the environment; or a model of political economy, a way of structuring the economic and political order of a society. For a collection of key texts that outline core ideas and evaluate critical debate, see Mol, P.J., Sonnenfeld, D.A., and Spaargaren, G. (eds.) 2010. *The ecological modernisation reader: Environmental reform in theory and practice*. London: Routledge. For a discussion of how ecomodernism has been applied in different countries, including ones outside Europe, see the special edition of *Environmental Politics* (volume 9, issue 1, 2000) on the topic of ecomodernism around the world.

## Bibliography

Adshead, M. 1996. Beyond Clientelism: Agricultural Networks in Ireland and the EU. *West European Politics* 19 (3): 583–608.

Barry, J. 2007. Towards a Model of Green Political Economy: From Ecological Modernisation to Economic Security. *International Journal of Green Economics* 1 (3/4): 466–464.

Bindi, M., and J.E. Olesen. 2011. The Responses of Agriculture in Europe to Climate Change. *Regional Environmental Change* 11.(Supple 1: S151–S158.

Burck, J., U. Hagan, N. Höhne, L. Nascimento, and C. Bals. 2020. *Results 2020. Climate Change Performance Index*. Bonn: Germanwatch, NewClimate Institute & Climate Action Network.

Christoff, P. 1996. Ecological Modernisation, Ecological Modernities. *Environmental Politics* 5 (3): 476–500.

Coghlan, O. 2007. Irish Climate-Change Policy from Kyoto to the Carbon Tax: A Two-Level Game Analysis of the Interplay of Knowledge and Power. *Irish Studies in International Affairs* 18: 131–153.

Cohen, M. 1998. Science and the Environment: Assessing Cultural Capacity for Ecological Modernisation. *Public Understanding of Science* 7: 149–167.

Collins, H., and R. Evans. 2017. *Why Democracies Need Science*. London: Polity.

Commins, P., and P. O'Hara. 1992. Restructuring in the Irish Agricultural Economy: Determinants, Processes, and Outcomes. In *Ireland and Poland: Comparative Perspectives*, ed. P. Clancy, M. Kelly, J. Wiater, and R. Zoltaniecki, 133–143. Dublin: UCD Press.

Devitt, C., and E. O'Neill. 2016. The Framing of Two Major Flood Episodes in the Irish Print News Media: Implications for Societal Adaptation to Living with Flood Risk. *Public Understanding of Science* 26 (7): 872–888.

Doran, P., and J. Barry. 2009. The Environmental Movement in Ireland. In *A Living Countryside? The Politics of Sustainable Development in Rural Ireland*, ed. J. McDonagh, T. Varley, and S. Shortall, 321–340. Farnham: Ashgate.

Dryzek, J.S. 2013. *The Politics of the Earth: Environmental Discourses*. Oxford: Oxford University Press.

Fahy, D. 2017a. Defining Objectivity, False Balance, and Advocacy in News Coverage of Climate Change. *Oxford Research Encyclopedia of Climate Science*. https://doi.org/10.1093/acrefore/9780190228620.013.345.

———. 2017b. The Fragile Culture of Science Journalism. In *Little Country, Big Talk: Science Communication in Ireland*, ed. B. Trench, P. Murphy, and D. Fahy, 37–60. Luton: Pantaneto Press.

Fox, E., and H. Rau. 2018. Climate Change Communication in Ireland. In *The Oxford Encyclopedia of Climate Change Communication*, 320–340. Oxford: Oxford University Press.

Government of Ireland. 2019. *Climate Action Plan to Tackle Climate Breakdown*. Dublin: Department of Communications, Climate Action & Environment.

Hulme, M. 2009. *Why We Disagree About Climate Change: Understanding Controversy, Inaction and Opportunity*. Cambridge: Cambridge University Press.

Irish Farmers Journal. 2019. Company Information. Available at: http://www.farmersjournal.ie/information-company/150940

Joint Committee on Climate Action. 2019. *Report of the Joint Committee on Climate Action*. Dublin: Houses of the Oireachtas.

Lanigan, G. 2017. Reducing Greenhouse Gas Emissions from Agriculture. *Teagasc*. Available at: https://www.teagasc.ie/publications/2017/reducing-greenhouse-gas-emissions-from-agriculture.php

Leonard, L. 2007. Environmentalism in Ireland: Ecological Modernisation Versus Populist Rural Sentiment. *Environmental Values* 16 (4): 463–483.

Little, C. 2017. Portrait of a Laggard? Environmental Politics and the Irish General Election of February 2016. *Environmental Politics* 26 (1): 183–188.

Little, C., and D. Torney. 2017. The Politics of Climate Change in Ireland: Symposium Introduction. *Irish Political Studies* 32 (2): 191–198.

Marsh, M. 2020. 'Poll of Polls: Fluctuation since 2015 but little real change' RTE. ie 14th Jan. 2020 https://www.rte.ie/news/election-2020/2020/0114/1107344-election-poll-of-polls/.

McGee, H., and K. O'Sullivan. 2019. Plan Aims to "Nudge" People into Change; Taoiseach Calls for Common-Ground Approach to Tackle Climate Change. *The Irish Times*, June 18.

McNair, B. 2011. *An Introduction to Political Communication*. 4th ed. Abingdon: Routledge.

Mol, A.P.J., and G. Spaargaren. 2000. Ecological Modernisation Theory in Debate: A Review. *Environmental Politics* 9 (1): 17–49.

O'Brien, M. 2017. *The Fourth Estate: Journalism in Twentieth-Century Ireland*. Manchester: Manchester University Press.

O'Malley, E. 1992. Industrialisation in Ireland. In *Ireland and Poland: Comparative Perspectives*, ed. P. Clancy, M. Kelly, J. Wiater, and R. Zoltaniecki, 105–117. Dublin: UCD Press.

Pepper, D. 1999. Ecological Modernisation or the 'Ideal Model' of Sustainable Development? Questions Prompted at Europe's Periphery. *Environmental Politics* 8 (4): 1–34.

Pielke, R., Jr. 2010. *The Climate Fix: What Scientists and Politicians Won't Tell You About Global Warming*. New York: Basic Books.

Robbins, D. 2018. Climate Change, Politics and the Press in Ireland 18 Oct 2018. Taylor & Francis.

———. 2019. Climate Change Frame Production: Perspectives from Government Ministers and Senior Media Strategists in Ireland. *Environmental Communication*. https://doi.org/10.1080/17524032.2019.1691620.

———. 2020. Climate Change Frame Production: Perspectives from Government Ministers and Senior Media Strategists in Ireland. *Environmental Communication* 14 (4): 509–521.

Shanagher, S., and P. Brereton. 2019. Pilgrim Hill: Alienated Farmers and Degraded Ecologies. *Capitalism, Nature, Socialism*. https://doi.org/10.1080/10455752.2019.1670710.

Share, P., H. Tovey, and M.P. Corcoran. 2007. *A Sociology of Ireland*. Dublin: Gill and McMillan.

Torney, D., and R. O'Gorman. 2019. A Laggard in Good Times and Bad? The Limited Impact of EU Membership on Ireland's Climate Change and Environmental Policy. *Irish Political Studies* 34 (4): 575–594.

Tovey, H. 1992. Environmentalism in Ireland. In *Ireland and Poland: Comparative Perspectives*, ed. P. Clancy, M. Kelly, J. Wiater, and R. Zoltaniecki. Dublin: UCD Press.

Trench, B. 2009. Representations of the Knowledge Economy: Irish Newspapers' Discourses on a Key Policy Idea. *Irish Communications Review* 11 (1): 3–20.

UN World Commission on Environment and Development. 1987. *Report of the World Commission on Environment and Development: Our Common Future.* Bruntland Report. New York: United Nations. https://sustainabledevelopment.un.org/content/documents/5987our-common-future.pdf.

Wagner, P., and D. Payne. 2017. Trends, Frames and Discourse Networks: Analysing the Coverage of Climate Change in Irish Newspapers. *Irish Journal of Sociology* 25 (1). https://doi.org/10.7227/IJS.0011.

Weale, A. 1992. *The New Politics of Pollution.* Manchester: Manchester University Press.

Yearley, S. 1995. From One Dependency to Another: The Political Economy of Science Policy in the Irish Republic in the Second Half of the Twentieth Century. *Science, Technology and Human Values* 20 (2): 171–196.

# Challenges and Potentials for Socio-Ecological Transformation: Considering Structural Aspects of Change

*Trish Morgan*

## Introduction

Ireland has been characterised as a climate laggard (O'Sullivan 2018) for its long-standing inaction on climate change. However, in order to situate Ireland's changing response to climate change, it needs to be recognised that its position is more insidious than that. It is a climate hypocrite, where the minister for climate action can on one hand claim to support the Schools Strike for Climate and yet, once a climate and biodiversity emergency has been declared in the Dáil in May 2019, grant a licence for off-shore fossil fuel extraction (Sargeant 2019). Indeed, such recent double-speak and 'smoke and mirrors' attitude to policy (Sargeant 2019) is not an isolated event. The former Taoiseach Enda Kenny was previously observed on the one hand making a commitment to act on climate change at the UN Climate Summit in New York in 2014, before asking the

T. Morgan (✉)
School of Communications, Dublin City University, Dublin, Ireland
e-mail: trish.morgan@dcu.ie





D. Robbins et al. (eds.), *Ireland and the Climate Crisis*, Palgrave Studies in Media and Environmental Communication,
https://doi.org/10.1007/978-3-030-47587-1_9

European Union (EU) within weeks for special treatment to reduce its commitment due to the large proportion of Irish GDP devoted to agriculture (Kenny 2014; RTÉ News 2014; Morgan 2018).

This chapter contends that such contradictions require a systemic and structural analysis in order to reveal the political economy, and indeed the political ecology in Ireland's response to climate change. Therefore, this chapter explores concepts of political economy, situating Ireland's position within macro-economic contexts. Next, it outlines key thought from the area of political ecology in order to contextualise Ireland's understanding of the nature-society relationship. Having discussed these conceptual areas, the chapter investigates the communication of environmental issues. This is done through a case study of the coverage of the fifth assessment report of the United Nations Intergovernmental Panel on Climate Change (IPCC AR5) by public sector television news and current affairs broadcasting. The case study finds that the amount and quality of coverage in Ireland needs to be considered alongside competing information streams that can dilute potential robust communication on environmental matters and, in turn, affect the ability to communicate important messages to audiences around transition to sustainability.

## Theories of Political Economy

For the purposes of understanding Ireland's response to climate change, this chapter suggests that a structural analysis of growth-based economic theories, and their tensions with ecological sustainability, is an appropriate lens through which to problematise the contradictions between the first-world consumption levels of Irish society and the urgent need to reduce unsustainable consumption.

### *Compounding Economic Growth*

I suggest that a consideration of compounding economic growth is of salience to our discussion of structural issues in the Irish context. For Harvey (2010, 2014), ideas of 'degrowth' or 'steady-state' economies are contradictory, and indeed impossible in an economic system such as that of Ireland. He saliently explains that a capitalist economy *requires* growth, noting that 'a zero-growth capitalist economy is a logical and exclusionary contradiction. It simply cannot exist' (Harvey 2014: 232). Therefore, if such an economy is *not* growing at an average of 3% per year, it is in crisis.

Thus, an imperative for the political domain is usually to provide a 'fix' to the crisis, either by deferring it temporally, by invoking various financial and fiscal measures, or by deferring it spatially, through economic expansion into other territories or, in the Irish case, encouraging and inward 'fix' through Foreign Direct Investment (FDI) (Castree and Gregory 2006).

When these ideas of growth, crisis, and fixes are applied to the Irish context, we can understand that any government is required to respond to economic stagnation through various measures to provide spatial and/or temporal fixes, to ensure that an upward growth trajectory is resumed. Therefore, in the Irish context, the implications of this basic economic imperative must be meaningfully considered before assuming that transition to a non-growth or differently growing system is a straightforward decision (Jackson 2009; Hamilton 2003).

Indeed, as a way of problematising these tensions, Steffen et al. (2006) first coined the phrase the 'great acceleration' to compare socio-economic trends of compounding economic growth, with those of earth system trends (Lane 2019). In a series of stark graphs, the impact of socio-economic trends is juxtaposed with stresses on the ecosystem (Steffen et al. 2006: 5–6), showing how compounding growth hits inflexion points, where such a growth curve 'rises very slowly for quite a while… and then starts to accelerate and by the end of the curve it becomes what mathematicians refer to as a singularity – it sails off into infinity' (Harvey 2014: 224). Indeed, Harvey categorises the compounding nature of growth as 'an extremely dangerous but largely unrecognised and unanalysed contradiction' (ibid.: 222).[1]

Furthermore, the 'growth fetish' (Hamilton 2003) of contemporary economies prioritises positive GDP over issues of diminishing well-being and deepening inequality. Discourses privileging growth over sustainability have been shown to erroneously conflate growth for growth's sake with happiness (Skidelsky and Skidelsky 2012; Sandel 2013; Jackson 2009). This is to the extent that a questioning of continued economic growth is seen as 'the act of lunatics, idealists, and revolutionaries' (Jackson 2009: 14).

However, a critical political economy approach reveals that the type of growth pursued in some economies over the last 35 or so years is no longer providing well-being for a majority of citizens. This is particularly the case in the UK and the US, where austerity measures have been enacted to ensure a return to the type of uneven growth characterised by these neoliberal economic policies. Such policies favour the deregulation of the private sector, minimal investment in public resources, and neglect of social protections and securities (Dorling 2015; Mendoza 2015). Concerns have arisen that these policies have caused declines in living standards and even life expectancy (Sayer 2016; Skidelsky and Skidelsky 2012).

Ireland, in its attempts to pursue an 'invisible ideology' (O'Callaghan et al. 2015: 33) of neoliberal economic policies, has also witnessed its own deepening inequality, with over 10,000 citizens homeless in 2019 (Focus Ireland 2019), as house prices have increased to 74% more than their lowest prices in 2012 (Irish Examiner 2019). Thus, we can see that in the Irish context, 'fixing' the economy back to growth has been a key priority in the last ten or so years, even if it means uneven distribution of the benefits of growth.

This has implications for environmental issues in any 'open' economies such as Ireland's, in that when (not if) another economic crisis occurs, there is a danger that progress on environmental issues may be sidelined in order to 'fix' the economy. This is borne out in 2019 figures from the Environmental Protection Agency revealing that Ireland is continually missing its targets on carbon emission reduction, as the economy returns to positive growth after a period of crisis (Environmental Protection Agency of Ireland 2019). The idea of fixes is discussed next, in order to provide a more in-depth analysis of how such responses work and how they need to be considered in the context of environmental crisis.

### *Spatial and Temporal Fixes*

The 'strange non-death of neoliberalism' (Crouch 2011, 2013) has shown that despite a crisis in growth-based economies in 2008, there has not been a systemic reshaping of the growth paradigm to which such economies subscribe. I suggest that the concept of the 'fix' is a useful one to analyse the Irish context. A fix is 'a general term that refers to many different forms of spatial reorganization and geographical expansion that serve to manage, at least for some time, crisis-tendencies inherent in accumulation' (Castree and Gregory 2006: 146). The concept of the fix suggests that when an economic crisis manifests in a region, capital is not necessarily lost, but merely moved around in space and/or time. This reveals that when there is uncertainty for the accumulation of capital in a region, it will find a spatial fix to its accumulation by moving to where it is easier to operate. While the spatial fix defers crisis by moving in geographic and territorial space, fixes can also occur in a temporal way through financial and credit instruments that can promise future dividends and profits, deferring lack of return on investment in capital (Harvey 2010; Castree and Gregory 2006). An aspect of the temporal fix is the use of consumer credit to offset the effects of declining real wages (Morgan 2018; Harvey 2010). Indeed, it can also be seen in the idea of environmental finance such as 'hurricanes future markets' (Smith 2007b: 777), where future profits can be guaranteed to capital through increasing instability of global climate.

It must be noted, however, that such fixes do not resolve the tendencies in growth-based economies to their vulnerability to crisis. They defer, manage, and move crises, in that in a certain region, citizens can be subject to austerity measures, while others can accumulate massive profits on 'distressed' loans. Indeed, certain cohorts can accumulate substantial wealth through the 'expropriation of occupational or public pensions and other funded future welfare entitlements for immediate profit' (Castree and Gregory 2006: 151). Furthermore, crises can open up new sites for development, with employment moved to other territories where it is cheaper to accumulate capital. Indeed, a function of a 'fix' is in how it 'displaces and defers contradictions both within a given economic space and/or political territory and beyond it. It also involves an internal as well as an external differentiation of winners and losers from a particular fix, linked to the uneven social and spatial distribution of benefits from a given fix and to its associated uneven development' (Castree and Gregory 2006: 163). Thus, we can conclude that both spatial and temporal fixes are therefore of salience to any territory in that they are part of the operating mechanism of contemporary economies.

In the Irish context, this 'fix' is seen in the temporal and spatial fixes associated with housing. The spatial fix in part involved the outflow or evaporation of capital that resulted in ghost estates (Kitchin et al. 2014). The temporal fix can be seen in the setting up of the National Assets Management Agency, a 'bad bank' that was seen as providing the temporal fix of bailing out by the Irish state of the non-performing financial sector (Byrne 2016), and the subsequent 'dumping of huge loan portfolios onto the market', including on to so-called vulture funds (Byrne 2016: 906).

## Theories of Political Ecology

### *The Nature-Society Relationship*

We have thus far problematised the relationship between economic growth and ecological sustainability. In the Irish context, it is necessary to ask how nature itself is understood in contemporary society. Ireland has a unique place on the global stage, as every year hundreds of buildings all over the world turn green on 17 March in celebration of Ireland and Irishness. Green is semiotically synonymous with environmental well-being and, therefore, associates Ireland with positive connotations of nature.

However, ideas of 'nature' are contested in contemporary societies. Therefore, what *can* be understood about the society-nature relationship is in flux and needs examination. For example, a common way to think of nature is *natural*, as is untouched by society and industry. However, aerial imagery of the Masai Serengeti, an ecologically homogenous landscape, reveals dramatically differing land use between the territories of Kenya and Tanzania (Robbins 2012). Thus, a so-called natural landscape can be variously affected by political and policy decisions, revealing that nature is not necessarily 'natural', but subject to political decisions that influence its overall health, biodiversity, and development.

From this political ecology perspective, we can thus infer that nature is subject to political and societal influences. This opens the discourse to an enquiry of other conceptual inferences made in society about nature. For example, a key trope in contemporary economies is in how nature is external to the economy. Thus, ecologically damaging practices, such as large-scale tree felling, can show up as a positive economic development as it adds to GDP (Hamilton 2003).

### *Where and When Is Nature?*

Furthermore, the asking of seemingly less complex questions such as *where is nature?* or *when is nature?* evokes telling answers in contemporary society (Castree 2014). In relation to the first question, nature is increasingly *away* from society or, indeed, more likely available to society in managed entities such as zoos and wildlife parks (ibid). The spread of urbanisation also increases 'the material production of environments [that are] necessarily impregnated with the mobilization of particular discourses and understandings (if not ideologies) of and about nature and the environment' (Henyen et al. 2006: 7). Indeed, for Ernston and Swyngedouw (2019: 13):

> While urbanization was always predicated on the extraction of material resources beyond the city proper and producing significant metabolic rifts, this has arguably reached a new scale and intensity in the present process of planetary and uneven urbanization.

This spatial analysis, that is, considering *where is nature?*, reveals a false separation between society and nature and an increasing distancing of nature from society, manifesting in contemporary urban political ecologies.

In relation to the second question, Castree (2014: 12) observes that the answer brings us to realise that nature is *increasingly in the past.* This temporal analysis throws into relief the idea that the great acceleration is, unless checked, a one-way arrow in terms of its destruction of nature. Through development, the great forests of Europe, the great Amazon rainforests, and the forests of Indonesia, which took hundreds of years to establish, are irrevocably damaged or eroded. The past 'lungs of the earth', as these great forests have been for society, cannot be restored to their mature function in the short time frame required of them to function as carbon sinks. Therefore, society needs to contend with the fact that this nature is in the past, and it is such because of societal practices that have favoured economic development and 'externalising' nature, over its care and maintenance.

This throws into sharp focus the centrality of environmentally appropriate practices, in this instance reforestation and rewilding policies, not as optional extras in development policy, but central to any plans for development globally and, indeed, in Ireland. For example, Fogarty (2017: 81) reports that out of 'fifty-eight types of protected habitat only five are considered to be in good shape', with over a thousand national heritage areas (NHAs) 'yet to receive official legal protection' (ibid.: 83). Yet, the 'ultra high-pressure upstream pipeline and inland refinery built by Shell E&P Ireland in the Corrib field off northwest Mayo are a prime example of high-risk technologies that are being trialled in the Irish semi-periphery (Deckard 2015: 164–165). Similarly, the redevelopment of Dublin port to enable the intensification of tourism from cruise ships neglects the environmental impacts of $CO_2$, NOx, and particulate pollution in this urban location (Byrne 2019). Such neglect of protected habitat, contrasted with the development of infrastructural projects that potentially damage the environment, is, in a political ecology frame, not neutral, but driven by varying priorities and agendas.

A key point is in how we conceptually understand the nature-society relationship both in the Irish context and beyond. An improvement in contemporary conceptual understanding of nature is through the concept of *metabolism* (Foster 2013). From this standpoint, the socio-ecological relationship is not one of separation but one where we have to acknowledge a profound interconnection with, and indeed dependency on, nature (Weintraub 2019; 2012). Just like the body's metabolism automatically keeps respiration, digestion, and temperature under control, the socio-ecological metabolism functions as a singularly integrated but complex system. However, just as the human metabolism responds to threats from viruses, parasites, or bacteria by temperature fluctuations and other unpleasant symptoms, the socio-ecological metabolism responds to agents that imbalance it, such as $CO_2$, plastics, or nitrates. Explained in this way, the concept of metabolism is a tangible way to understand the current 'rift' in the nature-society relationship (Foster 2013). When through ecologically unsustainable practices society exhausts resources and brings about the extinction of species, we can understand that we are also causing a rift in this metabolism, however separate those domains appear in contemporary society (Moore 2011; Foster 2013). Viewed in this light, issues of biodiversity loss become urgent issues of both planetary and societal health, and indeed survival. Thus, the concept of *metabolism* helps nuance how society can view its approach to the natural world.

### *Universal Nature/External Nature*

The work of Neil Smith has significantly contributed to the domain of political ecology and is, therefore, important to discuss. Smith contributed on two counts relevant to this chapter: (1) how nature is conceived under contemporary socio-economic arrangements and (2) the concept of the pr*oduction of nature* itself. I start with a brief overview of Smith's (2008) thesis on the 'ideology of nature'. This is to reveal the strengths of a political ecology approach to how nature is conceptualised in contemporary thought. This then lends itself to an analysis of how nature is produced under current economic arrangements.

For Smith, contemporary socio-economic arrangements have influenced how nature is perceived in society (Smith 2008: 10). Thus, the 'domination of nature' ideology, where nature is utilised for human development, is commonplace (ibid.). This increasing impact by economic forces on the natural world has implications for the socio-ecological relationship. While Smith acknowledges how concepts of nature can be complex and contradictory, his central contribution to debates on nature in contemporary contexts centres on a fundamental reduction of the complexity of the nature-society relationship to a dualism (Smith 2008: 11). This dualism is, on the one hand, between the idea of nature as '*external*, a thing, the realm of extra human objects and processes existing outside society' (ibid.: 11). On the other hand, it is also 'universal' (ibid.: 11) because it includes humans, and therefore in the contemporary setting 'ecological treatments of human society situate the human species as one among many in the totality of nature' (ibid.: 12). Smith observes that this is a contradictory dualism, in that, on the one hand, nature is external to human society when considered in some domains of knowledge, yet because humans are also conceived of as being part of nature, nature therefore has to be thought of as universal. As Smith notes, 'external and universal nature are not entirely reconcilable, for at the same time that nature is held to be external to human existence, it is simultaneously both external and internal' (ibid.: 12).

Therefore, we can ascertain that as long as the external conception of nature is prevalent in ecological discourses, there can be a justification for treating nature as separate from society and subject to domination, extraction, and use by contemporary economies, with little or no thought to ecological maintenance let alone restoration.

The second duality, that of universalism, is, for Smith, a vehicle for ideology (ibid.: 29). This can be mobilised 'to justify the conquest of external nature'. The effect of such ideology is, for Smith, 'one of conquest – or more accurately control' (ibid.). It is important for our discussion of political ecology to consider the systemic thought and assumptions that are made with respect to both nature and the economy. Smith's skill here is to place what has been justified as 'natural', such as there being no alternative to growth-based economies, into a frame where it becomes historically situated and therefore changeable. This leads to a de-naturalising of our contemporary assumptions around both nature and society and raises questions pertaining to how nature is produced under contemporary contexts. I therefore now turn to Smith's ideas on the production of nature.

### *Production of Nature*

Smith's production of nature thesis echoes the political economy perspective of Harvey (2010, 2014) in that it takes account of compounding economic growth and its impacts on the ecosystem. For Smith, capitalism is distinctive in that 'under dictate from the accumulation process, capitalism as a mode of production must expand continuously if it is to survive' (Smith 2008: 71). Thus, the evolving relationship between the economy and nature becomes problematic, Smith observing that 'to this end [growth], capital stalks the earth in search of material resources; nature becomes a *universal means of production* in the sense that it not only provides the subjects, objects, and instruments of production, but is also in its totality an appendage to the production process' (ibid.: 710).

For Smith, as long as these conditions exist, the societal relationship with nature cannot fundamentally shift from one based on the production of nature as an economic end. He cites ideas of 'green capitalism', wetland credits, debt-for-nature swaps, and the fields of environmental banking and environmental derivatives as examples of how environmental issues are treated as commodities (Smith 2007a: 17). Thus, considering the production of nature thesis sheds critical light on the paradox of the societal assumptions of nature as 'natural' and not subject to being produced. Smith's thesis reveals how nature is produced in contemporary societies in a multitude of ways. Furthermore, how nature is produced under 'green capitalism' is of concern in that it can act as a source of further accumulation, potentially valuing certain environmental issues over others. This is to the extent that as the system has expanded and required new markets to continue its accumulation, this 'explosion of ecological commodification and capitalization has significantly deepened the production of nature' (Smith 2007a: 25). This is to the point that 'it is the regulation and production of nature that threatens to penetrate "all the way down"' (ibid.). He notes that this incorporation within the financial and market system implies that 'the same credit system that supposedly protects a wetland or forest can lead to its destruction when the credit system itself collapses' (ibid.: 34).

Smith's production of nature thesis is of importance to this chapter as it indicates how resolving the societal tensions with nature are unlikely to come from the same system that has been shown to create environmental crises. Smith's work reveals how the system itself is contradictory and therefore unable to resolve its own relationship with nature, be that an externalising or universalising one.

## A Structural Analysis of Knowledge Production: Irish News Broadcasting During the IPCC AR5 Reports

Given that this chapter takes a structural approach to issues concerning climate responses in Ireland, it now turns to structural approaches to assessing the place of knowledge production in communicating issues of environmental concern to citizens. Given the focus of this chapter, we need to apply such perspectives to production of environmental knowledge. The field theory of Pierre Bourdieu takes account of this dimension.

Bourdieu considers a field to be 'a social arena within which struggles of manoeuvres take place over specific resources or stakes and access to them' (Jenkins 2002: 84). Such struggles relate to the 'social resources specific to that field, the intellectual field, the educational field, the economic field etc.' (Garnham and Williams 1980: 215). This means that both intangible *and* structural issues are at play, and they take place both within fields and between fields. This is of salience to us as it holds implications for the field of knowledge production, in that norms within fields, and also external factors, can influence what knowledge can be produced. Bourdieu does not consider what he terms the symbolic field, in which we can situate the field of knowledge production, to be any different to other fields, in that 'the symbolic field like all fields is a field of class struggle, and what is at stake is legitimizing or delegitimizing power' (ibid.).

The concept of the field has implications for the production of media knowledge. Bourdieu posits that a cultural field 'is a veritable social universe where, in accordance with its particular laws, there accumulates a particular form of capital and where relations of force of a particular type are exerted' (Bourdieu 1996: 164). In this case, the specific capital can be understood as either cultural or social capital, rather than exclusively material or financial capital (ibid.). Thus, discourses in these fields are influenced by both the hierarchical struggles at play within the field of knowledge production and also by the relationship of the field of knowledge production to the broader field of power structures. Therefore, within the hierarchy of a field, there can exist internal struggles, while the field itself can possess varying amounts of autonomy when viewed relative to the field of power.

For the purposes of this analysis, the relative autonomy of the journalistic field is worth considering, as it is a frequent field of analysis for the area of environmental communication. Applying Bourdieu's field theory acknowledges that the journalistic field has its own internal logics, norms, and hierarchies (Preston 2009). At the same time, it is also situated in a hierarchy of capital accumulation and the power dynamics that such structures bring about. Bourdieu stresses the importance of considering the field of knowledge production in this manner, as it helps explain certain strategies to employed members of the field of knowledge production, for example, journalists. Indeed, he uses journalism in an annotated case study to apply his field theory (Bourdieu 1996: 124; Fig. 1).

For Bourdieu, the overarching social space (blue) sets the backdrop for the field of power (purple) and, in turn, within that various other fields such as the field of knowledge or, in Bourdieu's terms, cultural production operate (green). In turn, Bourdieu splits this field into those of large-scale production, such as that of journalism, and small-scale production, such as that of art. This analysis provides Bourdieu with the facility to then compare various levels of capitals within those domains; the field of journalism is characterised by a high level of economic capital (CE+), but a low level of cultural capital (CC−) and autonomy of expression (AUTON−). By contrast, the domain of art possesses little economic capital (CE−) but much cultural capital (CC+) and high levels of autonomy (AUTON+).

We can, therefore, observe that the relative autonomy of various domains of knowledge production is important in considering environmental discourse as a whole. For Bourdieu, journalism has a relative autonomy, but the industrialisation or commodification of this field brings in a power dynamic that potentially offsets or dilutes that autonomy. Indeed, while enjoying a weak form of relative autonomy, the journalistic field of knowledge production contends with the political and social science fields for legitimacy (Bourdieu in Benson 2005: 29). Bourdieu therefore argues that to fully understand the field of journalism, one needs to look to financial and advertising interests and their interaction with journalistic products. Thus, the domain of journalism is revealed to be subject to myriad external pressures on its autonomy and, therefore, its ability to report on issues in a way that is unaffected by these pressures.

This chapter therefore suggests that a structural analysis such as that of Bourdieu can help situate the media as a domain of symbolic production

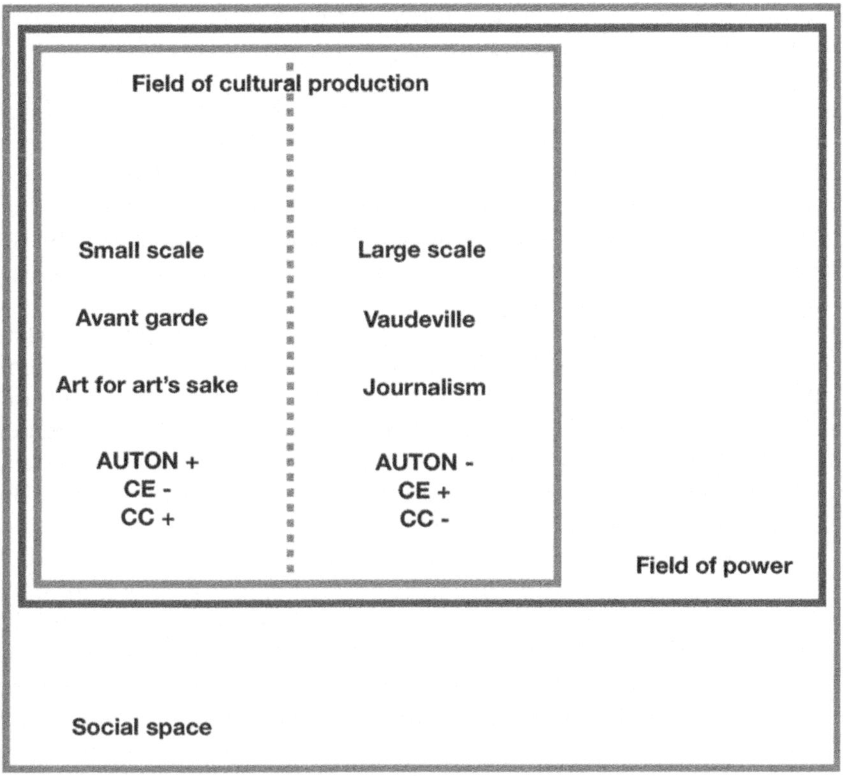


**Fig. 1** Bourdieu's field theory (Bourdieu 1996: 124), re-annotated by author

that cannot be isolated from its economic contexts and pressures. It is with this in mind that the chapter briefly presents a case study of the Irish television broadcast coverage of the 5th assessment report of the Intergovernmental Panel on Climate Change (IPCC AR5).

### *Case Study*

The case study summarised here, and available in full in Morgan (2017), provides an analysis of a selection of public service TV news and current affairs broadcasts in Ireland at the time of the release of the 5th assessment report of the Intergovernmental Panel on Climate Change (IPCC AR5).

The content for the analysis comprised the main RTÉ News broadcasts, recorded in four two-week periods across 2013 and 2014, which covered the four release dates of the IPCC reports. It included broadcasts one week either side of the release of the reports. A total of 773 news stories were sampled for their thematic content during these periods. As this work was concerned with the overall structure of the content, rather than a discourse analysis or framing of the individual stories themselves, the number of advertisements—618—within the broadcast was also counted. This also takes account of Bourdieu's findings regarding the external pressures on the autonomy of journalism.

The case study revealed four key findings. The first finding was in what the report terms the 'bundling' of journalistic and advertising content (Morgan 2017). This means that when an audience views the broadcast, they also are shown advertising content, thus giving mixed messages around consumption practices and an overall sense of 'business as usual'. This also exemplifies Bourdieu's observations on the external pressures on the field of journalism from areas of commercial activity.

The second finding was that coverage was episodic rather than thematic. Considering whether the environmental stories are covered in an episodic or thematic way over time (Iyengar 1991) can provide insight into whether the story is treated only on a particular day, or whether it is broadcast on subsequent days, and potentially from multiple perspectives. As shown in Fig. 2, day 8, marked in red, corresponds with the release

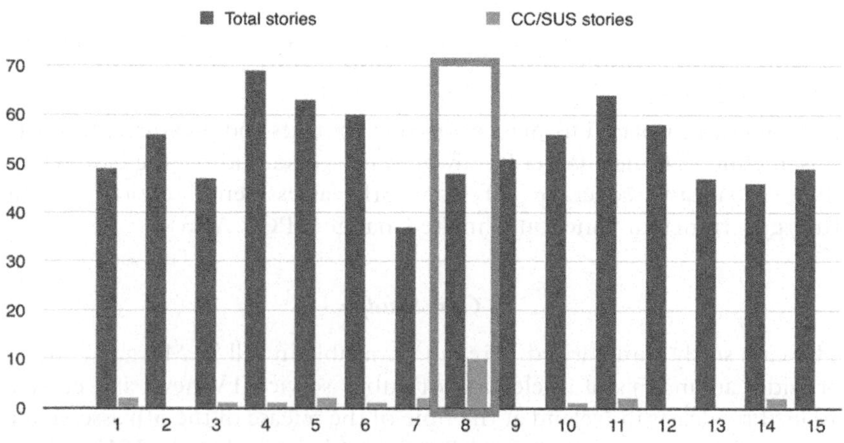


**Fig. 2** Aggregate coverage pattern across all time periods

of the IPCC report and, across the four reporting periods, shows that coverage is higher on the report release date than on other days. However, the aggregate data shows that throughout the four reporting periods, no environmental coverage occurred on day 9. This indicates that the stories related to the IPCC reports were treated on the day of their release, but not followed through on in subsequent days.

The third finding was in relation to content within the news broadcast itself, which revealed ironies and contradictions in messages. A key example was on the release date of the Working Group two (WGII) report, in which the effects of climate change were described as 'severe, pervasive and irreversible' (IPCC 2014: 14). RTÉ covered this well across several stories, and it was the top story on the broadcast. However, when the first advertising break was announced, it cut to the break with a 'feel good' story about the launch of nine new air routes from Shannon, and a return of daily long-haul services, arguably celebrating the unsustainable consumption of space (Morgan 2017: 32).

The fourth finding was around the seasonality and sophistication of advertising. The report revealed how the second and third IPCC reports around March-April coincided with Easter, and this corresponded with 45% food advertising in the same broadcast as the IPCC reports were announced. Upon analysing the advertisements more closely, luxury goods and chocolates were heavily advertised for the season of Easter. Likewise, the fourth report occurred around Halloween, a time that had 43% of advertisements devoted to food, with the advertisements showing treats and bulk sweets on offer from large supermarkets. In the area of transport, the variance was also significant, ranging from a low of 5% around March/April to a high of 17% in November, coinciding with a time when car manufacturers were advertising new model cars that would be available with the following year's registration, that is, in November 2014, manufacturers were advertising how customers could avail of a 2015 registration car. This reveals a seasonal variability in advertising content, indicating shifting and sophisticated messages around consumption.

We can thus infer from this case study that when viewed in terms of fields of knowledge production, the field of broadcast journalistic news production by necessity in the Irish context interacts with the domain of advertising. For Bourdieu, this constitutes an external pressure on the autonomy of the journalistic content. This suggests that irrespective of the quality of individual news stories on environmental issues, the overall 'bundle' of content received by the audience is not only one of mixed messages in the journalistic content itself, but compounded by the advertising content that celebrates consumption. Overall, the interaction of these two fields in the broadcast leads to an overall impression that while there are serious environmental issues, society can maintain a business-as-usual approach to consumption.

## Discussion and Conclusions

This chapter argues that applying political economy theories, including those of fixes in the Irish context, reveals particularities with respect to the response to crisis and also the unintended environmental consequences of those moves. Furthermore, the application of political ecology, such as the production of nature thesis, can also shed light on the interconnectedness of economic and environmental policies and practices in the Irish context. Finally, in terms of communicating environmental messages to publics, a structural approach to fields of knowledge production can assist understanding of the relative autonomy of various fields to communicate the radical action necessary for a critique of current policies and transition towards a more sustainable society.

In the contemporary Irish context, these three areas—political economy, political ecology, and structural approaches to knowledge production—can provide a lens through which to view Ireland's response to climate change. Broadly, Ireland is seen as a tax haven and thus a destination for 'fixes'. More specifically, Ireland is also seen as pursuing growth strategies in carbon-intensive sectors, such as agriculture. Such aggressive growth policies in this sector reveal a political economy concerned with unsustainable growth and intensification over diversification and sustainability of the sector.

However, these practices come with a political ecological impact also, with the attempted (and failed) rollback of the Wildlife Act to allow for more intensive hedge cutting, and the policy of planting Sitka spruce forestry, which is detrimental to biodiversity, and also critiqued for its unsustainability. This shows a tendency to privilege economic fixes over ecological ones, revealing how Ireland's response to climate change is both political economic and political ecological.

Finally, with these understandings, we can critique the communication of these key issues by taking a structural, field-based approach to knowledge production. It is assumed that more or better journalistic coverage on environmental issues is helpful. While this is to an extent valid, the case study summarised here questions the ability for salient messages to be communicated when internal journalistic ironies are present within broadcasting. Furthermore, treating Irish broadcast news as a 'bundle' of content shows that the nexus of the fields of journalistic and advertising production can send an unhelpful overall 'business-as-usual' consumption message to audiences.

To conclude, policymakers and stakeholders require pragmatic, measurable solutions to environmental challenges in the Irish context. However, this chapter has shown how macro-level factors can influence situated stakeholders in terms of the socio-ecological challenges faced in contemporary Irish society. Therefore, it contends that a consideration to these structural challenges can potentially inform policy through more systemic consideration of interconnected environmental issues, from production, circulation, and consumption practices in the Irish context.

## Note

1. See the work of the International Geosphere-Biosphere Programme for a salient collation of the socio-economic and earth system trends: http://www.igbp.net/globalchange/greatacceleration.4.1b8ae20512db692f2a680001630.html. Last accessed 4 June 2019.

## Bibliography

Benson, R. 2005. *Bourdieu and the Journalistic Field*. Cambridge: Polity.

Bourdieu, P. 1996. *The Field of Cultural Production: Essays on Art and Literature*. Cambridge: Polity.

Byrne, M. 2016. Entrepreneurial Urbanism After the Crisis: Ireland's 'Bad Bank' and the Redevelopment of Dublin's Docklands. *Antipode* 48 (4): 899–918.

———. 2019. '*Cruise Tourism in Dublin City*', *Political ecologies of Ireland* [online]. Available at: https://politicalecologytcd.wordpress.com/2019/05/14/cruise-tourism-in-dublin-city/. Accessed 4 Feb 2020.

Castree, N. 2014. *Making Sense of Nature*. Abingdon: Routledge.

Castree, N., and D. Gregory, eds. 2006. *David Harvey A Critical Reader*. Oxford: Blackwell.

Crouch, C. 2011. *The strange non-death of neoliberalism*. Cambridge: Polity.

———. 2013. *Making Capitalism Fit for Society*. Cambridge: Polity.

Deckard, S. 2015. World-Ecology and Ireland: The Neoliberal Ecological Regime. *Journal of World-Systems Research* 22 (1): 145–176.

Dorling, D. 2015. *Injustice – Why Social Inequality Still Exists*. Bristol: Policy Press.

Environmental Protection Agency of Ireland. 2019. *Ireland's Greenhouse Gas Emissions Projections 2018–2040*. Dublin: Environmental Protection Agency of Ireland.

Ernston, H., and E. Swyngedouw. 2019. *Urban Political Ecology in the Anthropo-Obscene: Interruptions and Possibilities*. Abingdon: Routledge.

Focus Ireland. 2019. About Homelessness. Available at https://www.focusireland.ie/resource-hub/about-homelessness/. Accessed 24 Oct 2019.

Fogarty, P. 2017. *Whittled Away: Ireland's Vanishing Nature*. Cork: The Collins Press.

Foster, J.B. 2013. Marx and the Rift in the Universal Metabolism of Nature. *Monthly Review* 65 (7): 1–20.

Garnham, N., and R. Williams. 1980. Pierre Bourdieu and the Sociology of Culture: An Introduction. *Media, Culture & Society* 2 (3): 209–223.

Gheibi, M., et al. 2018. Controlling Air Pollution in a City: A Perspective from SOAR-PESTLE Analysis. *Integrated Environmental Assessment and Management* 14 (4): 480–488.

Hamilton, C. 2003. *Growth Fetish*. Crows Nest: Allen & Unwin.

Harvey, D. 2010. *The enigma of capital*. Oxford: Oxford University Press.

———. 2014. *Seventeen Contradictions and the End of Capitalism*. New York: Oxford University Press.

Henyen, N., M. Kaika, and E. Swyngedouw. 2006. *In the Nature of Cities: Urban Political Ecology and the Politics of Urban Metabolism*, Routledge. Abingdon.

IPCC (Intergovernmental Panel on Climate Change). 2014. Summary for Policymakers. In *Climate Change 2014: Impacts, Adaptation, and Vulnerability. Part A: Global and Sectoral Aspects. Contribution of Working Group II to the Fifth Assessment Report of the Intergovernmental Panel on Climate Change*, ed. C.B. Field, V.R. Barros, D.J. Dokken, et al., 1–32. Cambridge: Cambridge University Press.

Irish Examiner. 2019. *House Prices Jumped by Almost 3% in 2019 – Report*. Available at: https://www.irishexaminer.com/breakingnews/ireland/house-prices-jumped-by-almost-3-in-2019-report-914441.html. Accessed 3 Oct 2019.

Irish Times. 2019. *Nissan to Review Sunderland Production in No Deal Brexit*. Available at https://www.irishtimes.com/business/manufacturing/nissan-to-review-sunderland-production-in-no-deal-brexit-1.4036735. Accessed 3 Oct 2019.

Iyengar, S. 1991. *Is Anyone Responsible?* London: University of Chicago Press.

Jackson, T. 2009. *Prosperity Without Growth : Economics for a Finite Planet*. London: Earthscan.

Jenkins, R. 2002. *Pierre Bourdieu*. Rev. ed. London: Routledge.

Kenny, E. 2014. Address by the Taoiseach to UN Secretary General's Climate Change Summit, New York, September 23. Available at: https://www.rte.ie/

documents/news/kennyclimatespeech.pdf; https://www.dfa.ie/irish-embassy/usa/news-and-events/2014/. Accessed 4 Jun 2019.

Kitchin, R., C. O'Callaghan, and J. Gleeson. 2014. The New Ruins of Ireland? Unfinished Estates in the Post-Celtic Tiger Era. *International Journal of Urban and Regional Research* 38 (3): 1069–1080.

Lane, R. 2019. The American Anthropocene – Economic Scarcity and Growth During the Great Acceleration. *Geoforum* 99: 11–21.

Mendoza, K.A. 2015. *Austerity – The Demolition of the Welfare State and the Rise of the Zombie Economy*. Oxford: New Internationalist Publications.

Moore, J.W. 2011. Ecology, Capital, and the Nature of Our Times: Accumulation and Crisis in the Capitalist World-Ecology. *Journal of World-Systems Research* 17 (1): 108–147.

Morgan, T. 2017. *Going Green Digitally? Environmental Crisis, Consumption Patterns, and the Evolving Role of Media*. Report prepared for the Environmental Protection Agency of Ireland (2013-SD-FS-1). Available at: https://www.epa.ie/pubs/reports/research/econ/research215.html

———. 2018. The Techno-Finance Fix: A Critical Analysis of International and Regional Environmental Policy Documents and Their Implications for Planning. *Progress in Planning* 119: 1–29. https://doi.org/10.1016/j.progress.2016.06.001.

O'Callaghan, C., et al. 2015. Topologies and Topographies of Ireland's Neoliberal Crisis. *Space and Polity* 19 (1): 31–46.

O'Sullivan, K. 2018. 'Extremely Concerning' Ireland Labelled a 'Laggard' Over Climate Action – An Taisce. *The Irish Times*. Available at: https://www.irishtimes.com/news/ireland/irish-news/extremely-concerning-ireland-labelled-a-laggard-over-climate-action-an-taisce-1.3534519. Accessed 3 Dec 2019.

Preston, P. 2009. *Making the News*. London: Routledge.

Robbins, P. 2012. *Political Ecology*. Oxford: Wiley Blackwell.

RTÉ News. 2014. *Multinationals Urged to Consider Ireland*. Available at: http://www.rte.ie/news/2014/1016/652658-asia-europe-summit/. Accessed 4 Jun 2019.

———. 2019. *Ireland Exceeds Annual EU Emissions Target by 5m Tonnes*. Available at https://www.rte.ie/news/2019/1024/1085284-epa-emissions-target/. Accessed 24 Oct 2019.

Sandel, M. 2013. *What Money Can't Buy*. London: Penguin Books.

Sargeant, N. 2019. New Drilling License Dhows Government's 'Smoke and Mirrors' Climate Policy. Available at: https://greennews.ie/new-drilling-gov-smoke-mirrors/. Accessed 4 Jun 2019.

Sayer, A. 2016. *Why We Can't Afford the Rich*. Bristol: Policy Press.

Shannon LNG. 2019. Available at www.shannonlng.ie. Accessed 24 Oct 2019.

Skidelsky, R., and E. Skidelsky. 2012. *How Much is Enough?* London: Penguin Books.

Smith, N. 2007a. Nature as Accumulation Strategy. *Socialist register* 2007: 16.

———. 2007b. Disastrous Accumulation. *South Atlantic Quarterly* 106 (4): 769–787.

———. 2008. *Uneven Development*. Athens, Georgia: University of Georgia Press.

Sridhar, R., et al. 2016. A Political, Economic, Social, Technological, Legal and Environmental (PESTLE) Approach for Assessment of Coastal Zone Management Practice in India. *International Review of Public Administration* 21: 211–232.

Steffen, W., et al. 2006. *Global Change and the Earth System*. Berlin: Springer Science & Business Media.

Weintraub, L. 2012. *To Life! Eco Art in Pursuit of a Sustainable Planet*. Berkeley: University of California Press.

———. 2019. *What's Next? Eco Materialism and Contemporary Art*. Bristol: Intellect.

# Climate Change and the Media

*David Robbins*

## Introduction

On June 23, 1988, NASA scientist James Hansen testified to a committee of the US Congress that he was '99% certain' that the rise in global temperatures in the 1980s was caused by human activity. The story was reported on the front page of the *New York Times* the next day (Shabecoff 1988), and climate change began its 'career' as a major media topic. Initially, there was a presumption that climate change would be resolved much in the way the ozone depletion crisis had been dealt with the previous year: this is, science drawing attention to a pressing problem and politicians responding with appropriate policy measures (see Ungar 2000, for a comparison between the two issues). Of course, it turned out to be a lot more complicated than that. Within a short time, scientists had ceased to be the major media source in climate change coverage, having been replaced by politicians, government officials, and industry figures (Carvalho 2007). In the 30 or so years since Hansen's testimony, effective policy responses remain elusive and, while coverage of climate change has

D. Robbins (✉)
School of Communications, Dublin City University, Dublin, Ireland
e-mail: david.robbins@dcu.ie



D. Robbins et al. (eds.), *Ireland and the Climate Crisis*, Palgrave Studies in Media and Environmental Communication,
https://doi.org/10.1007/978-3-030-47587-1_10

increased, the media's role in shaping social responses to the issue has come under increasing scrutiny.

Research into media coverage of climate change is a growing and diverse academic field (Schäfer and Schlichting 2014) and is based mainly on three effects of news media:

- to promote an informed public, especially in relation to scientific issues (Nelkin 1987; Wilson 1995);
- to influence public perceptions on the salience of a particular issue (McCombs 2004);
- and to frame the ways in which the public conceives of such issues (Goffman 1974; de Blasio and Sorice 2013).

Thus, for researchers interested in social responses to the challenges presented by climate change, levels of media coverage and the content of media texts can, to paraphrase Cohen (1963: 13), provide important indicators of what the public thinks about and how they think about it.

These three media effects—informing the public, agenda-setting, and framing—have each influenced different strands of research on media coverage of climate change. Early studies concentrated on whether the news media were fulfilling their normative role of accurately informing readers and examined whether media representations presented climate change as a settled—as opposed to contested—issue. Studies also examined the journalistic norm of balance to establish whether the media included sceptics or deniers in pursuit of 'balanced' coverage (Boykoff and Boykoff 2004, 2007). Later studies examined sceptic/denier presence in coverage (Painter and Ashe 2012; Painter and Gavin 2015).

In the agenda-setting strand of research, scholars became interested in coverage levels and the influences which hampered or encouraged media attention for the issue. The extent of coverage has been examined for many territories (e.g. see Schmidt et al. 2013, for a comparative study of coverage levels in 27 countries), but studies have tended to focus on the US and the UK, with a relative lack of comparative studies, and few studies which examine more than one media type (Schäfer and Schlichting 2014).

The way in which climate change is framed in the media has also been extensively studied, and a broad typology of frames has emerged from this scholarship: settled science, unsettled/contested science, disaster, opportunity, political/ideological, morality/ethics, health, and security (Nisbet 2009; Dirikx and Gelders 2010; O'Neill et al. 2015b).

From these three strands of research, we can make certain statements about media coverage of climate change: it has steadily increased since the 1980s; it suffers dramatic peaks and troughs, rising at the times of international climate conferences and then fading away; global coverage suffered a dramatic decline in late 2009 from which it is only now recovering; coverage is often framed politically (i.e. in terms of political advantage of one party/actor over another) (Boykoff et al. 2019), and it often features discourses related to ecomodernism, suggesting that solutions to climate change can be found within existing economic and political systems and that market mechanisms, technological fixes, and incentives can be used as policy levers to bring about the necessary emissions reductions.

More recent research has moved from treating climate change as a homogenous topic to analysing particular aspects, contributory factors, and impacts of climate breakdown. Studies have examined social media coverage of individual UN conferences (O'Neill et al. 2015a), flooding (Devitt and Neill 2017), air pollution (Duan and Takahashi 2017), representations of coal (Bacon and Nash 2012), carbon (Feldpausch-Parker et al. 2015; McNally 2015), and fracking (Jaspal and Nerlich 2014). Disruption and disaggregation have become features of the media environment; likewise, the topic of media coverage of climate change has become fragmented into smaller constituent elements.

## Social Media Coverage of Climate Change

Most studies of media coverage of climate change examine print media; 'coverage' on social media is a relatively neglected area, and such studies that exist tend to focus on Twitter. Research into representations of climate change on social media focus on three areas: publics (usually an analysis of the most influential accounts posting at the time of a climate change event), themes (discourses clustered around hashtags and @ mentions), and professional communication (analysis of campaigns and activism) (Pearce et al. 2019). A surprising feature of Twitter content relating to climate change is the extent to which mainstream media sources dominate communication on the platform. For instance, a study of Twitter activity at the time of the release of the 2013 Fifth Assessment Report (AR5) of the Intergovernmental Panel on Climate Change (IPCC) found that 78 of the 100 most frequently occurring domain names belonged to mainstream media outlets (Newman 2017), while a study of a random week on Twitter (not around a major news event) found that 67% of links in posts were to 'professional news organisations' (Veltri and Atanasova 2017: 733). Twitter activity also increases dramatically in response to the announcement of temperature anomalies (Kirilenko et al. 2015).

Social media activity relating to climate change is focused on the 'settled science' frame (O'Neill et al. 2015a), with prominent users engaged in defending climate science against attack. It is worth noting that concentrating on the validity of scientific findings has the effect of prolonging, rather than discouraging, debates concerning the credibility of climate science (Demeritt 2001), while downplaying more political debates about adaptation and mitigation policies. There is mixed evidence regarding whether communication concerning climate change on social media takes place in 'echo chambers', contributing to polarisation on the issue. One influential study finds that Twitter climate change discussions are 'characterised by strong attitude-based homophily and wide-spread segregation of users' into like-minded communities (Williams et al. 2015: 135), while other researchers have found 'grounds for cautious optimism' regarding the possibilities for open communications within 'mixed attitude' online communities (Pearce et al. 2014: 8). The literature on social media communications on climate change also finds that, while activists and sceptics seldom engage with each other, activists are more active on Twitter, with 'greater potential audience reach' (Pearce et al. 2019: 6).

Research into social media activity around climate change is a relatively new area for media scholars, and many gaps remain. Platforms such as Facebook, YouTube, Instagram, and non-English language platforms with extensive reach, such as Sina Weibo and Baidu Tieba (both Chinese), have been neglected up to now. Visual communication about climate change on social media platforms is also under-explored. Researchers have been attracted by access to large datasets made available by some social media companies, and there has been a lack of qualitative studies. There have been calls for research to focus more deeply on the activities of public figures or to explore differing climate imaginaries by analysing 'post content and meaning making through interactions' (Ibid.: 10).

## Media Coverage of Climate Change in Ireland

Ireland is a somewhat neglected territory when it comes to research into media coverage of climate change. The country has featured in some longitudinal and cross-country studies (Schmidt et al. 2013), and there has been research into media representations of flooding (Devitt and

Neill 2017), decarbonisation and low-carbon transition (McNally 2015), framing of the Papal encyclical on climate change in the Irish media (Robbins 2016), and a broad study of media attention levels and frames in Irish media from 2007 to 2016 (Robbins 2019). An overview of the scholarship of media coverage of climate change in Ireland has also been provided by an entry on Ireland in the Oxford Research Encyclopaedia of Climate Science (Fox and Rau 2016). The authors survey some of the literature mentioned above, concluding that (i) coverage of climate change has steadily increased since 1997; (ii) there was a steep 'post-crash drop' in late 2009; (iii) the *Irish Times* has a 'track record of providing the most coverage', and (iv) that coverage is largely driven by events (2016: 8). There is a striking lack of research into online and social media, while the only study on broadcast media was limited in scope, concluding the coverage by the state broadcaster RTÉ was 'infrequent, sporadic and clustered around a small number of topical areas' (Cullinane and Watson 2014: 19).

It would appear that Ireland's media are quite typical with regard to how and how often they cover climate change. Irish print media coverage exhibits more or less the same peaks and troughs as many other territories; levels of coverage rise unsteadily before suffering a steep decline in late 2009 following the collapse of COP15 in Copenhagen and the onset of the global financial crisis. Coverage of climate change accounts for 0.843% of total news coverage (Robbins 2019), a finding that is not dissimilar to that of Andreas Schmidt and colleagues in their survey of coverage in 27 countries (Schmidt et al. 2013).[1] The coverage is episodic, rising sharply around Conference of the Parties (COPs) and United Nations Framework Convention on Climate Change (UNFCCC) reports before fading almost to invisibility. Journalistic content focuses on party politics and jockeying for position, presenting climate change as a form of political contest in which there are winners and losers and in which various actors vie for advantage. The media do not portray climate change as an issue requiring radical changes to existing political, economic, or agricultural systems; they are political with a small 'p' and climate change coverage is not interested in 'the political' (Mouffe 2005; Carvalho et al. 2016). The Irish print media present a narrow, ideological view of climate change, one that is supportive of the country's political and economic elites (Wagner and Payne 2015). Solutions to the climate crisis are instead presented as possible from within existing economic, political, and social systems. The discourse of ecomodernism is strongly present in Irish coverage, whereby efficiencies, market-based incentives, and technological advances are emphasised as appropriate and adequate responses (McNally 2015; Robbins 2019).

Thus, media coverage of climate change in Ireland has much in common with the coverage found in other territories. Yet there are several aspects of this coverage that are unique to Ireland. The first relates to Ireland's media system. The scholarship of media systems places Ireland in the North Atlantic/Liberal model thanks to its medium newspaper circulation, neutral commercial press, and strong professionalisation within the journalism sector (Hallin and Mancini 2004). This categorisation suggests that Ireland's media are concerned with the differentiation between news and comment and are inclined merely to report on events rather than analyse/contextualise them or campaign for change, in contrast to the more polemical and politically engaged journalism in the European tradition (Chalaby 1996). This passivity in the face of a complex social problem was also evident in the coverage of Ireland's banking and housing crises (Mercille 2014a, b).

The second involves political and economic events in Ireland since 2007, namely, the economic crash, and the entry into government of the Green Party. Ireland provides a vivid example of the 'crowding out' effect whereby environmental topics are pushed down the media agenda to make way for news of financial events and armed conflict (Djerf-Pierre 2012). Print media coverage of climate change declined steeply in late 2009 and early 2010; however, the effect of the financial crisis on media coverage is complex: the timelines do not exactly correspond, as the financial crisis became evident in Ireland in late 2008, while the coverage of climate change does not exhibit its decline until a year later. Further research is required to understand whether the December 2009 decline was the result of the 'crowding out' effect or rather the impact of the failure of the Copenhagen COP and the disillusionment that set in among policymakers and journalists in the wake of the conference (Lyytimäki 2011).

The entry of the Green Party into government in 2007, and its period in power up to 2011, also had an impact on climate change coverage, due to the prominence of the issue in the party's policy agenda. Reporters, editors, and advisors working for Green Party ministers report an increase in coverage due to ministerial activities and initiatives, yet the data for media attention is more complex (Robbins 2019). It shows an increase in coverage mentioning both climate change and the Green Party in the summer of 2007 (election and government formation talks) and in

December 2010 and January 2011 (dissolution of coalition government), suggesting that political events involving the Greens increase media interest in climate change. However, at other times of elevated media coverage of climate change (COP15 and COP21, for instance), the Green Party is not included in reports. This suggests that the link between climate change and the Green Party in the media's approach is tenuous and non-reciprocal: when the Greens are in the news, they bring climate change with them; when climate change is in the news, the Greens are not included in the coverage (Ibid.).

## Sceptics, Irish-Style

Another peculiarity of Irish coverage is the representation of sceptic/denier views. The 'contested science' frame was found to be present in 11% of media coverage of climate change from 2007 to 2016 and was particularly evident in the *Irish Daily Mail* (Robbins 2019). In other territories featuring sceptic/denier viewpoints (almost exclusively the UK, the US, and Australia) (Antilla 2005; McKewon 2012; Painter and Ashe 2012), arguments take the form outlined in Rahmstorf's typology: trend, attribution, and impact sceptics (Rahmstorf 2004). Irish coverage features 'climate dismissives'—journalists, mostly elderly, male opinion columnists, who do not engage with the finding of climate science, but rather deride those advocating climate action as overly earnest and sanctimonious. Another feature of Irish coverage is the seemingly 'honest broker', a journalist adopting the pose of an open-minded 'everyman' waiting for conclusive evidence of a changing climate (Robbins 2019).

A further complicating factor in Ireland's coverage is the presence of a powerful lobbying organisation advocating for the agriculture sector. Agriculture contributes 33.1% of Ireland's total greenhouse gas (GHG) emissions (Government of Ireland 2017: 80), compared to an EU-28 average of 10% and a global average of 20% (IPCC AR5). Only New Zealand, with 48% (Government of New Zealand 2019), has a higher agriculture component. Coverage of climate change which mentions agriculture contains arguments that (i) carbon accounting does not allow sufficiently for carbon sinks; (ii) emissions reduction targets and agriculture production targets are incompatible; (iii) more carbon-intensive systems would replace Irish systems were Ireland to reduce production, and, more recently, that (iv) emissions reductions efforts should focus on the transport and energy sectors.

The recent emergence of a climate contrarian organisation in Ireland entitled the Irish Climate Science Forum has challenged a broad social and political consensus on the anthropogenic origins of recent climate change (Gibbons 2017). The group has, at the time of writing, hosted two closed meetings at which noted climate deniers (Richard Lindzen and William Happer) addressed the audience. Lindzen was the subject of an open letter from MIT scientists in which they dissociated themselves from Lindzen's appeal to President Trump urging the US to withdraw from the Paris climate change agreement (Hirji 2017), while Happer, among those being considered for the post of science advisor to the White House, has espoused a range of denier positions (Readfearn 2017). Both Happer and Lindzen questioned the contribution of methane to global warming. The meetings were covered by the media (even though journalists were prevented from attending) and the claim that methane was contributing less to the warming effect than previously thought was dismissed as 'balderdash' by eminent climate scientist John Sweeney (Boucher-Hayes 2017).

Although denier arguments are mostly relegated to non-mainstream venues such as the Irish Climate Science Forum, they have surfaced recently on online platforms such as Twitter and YouTube, mostly as the result of the activities of Gemma O'Doherty, a former journalist and provocateur, predominantly on issues relating to immigration, race, and religion (Smith 2019). O'Doherty supports various climate change conspiracy theories: climate change is a vast scientific hoax; climate change is not happening; climate change is a means whereby drastic lifestyle changes will be imposed by Leftists. She has, at the time of writing, 33,300 followers on Twitter and had 26,000 subscribers to her channel on YouTube before the platform removed her account for repeated breaches of its hate speech policy (O'Brien 2019).

The foundation of the Climate Science Forum, and the significant following for O'Doherty, suggests that there is a section of Irish society willing to entertain climate sceptic discourses. These discourses have been kept to the margins, and most mainstream and professional media coverage of climate change sceptics is assiduous in alerting the public to mischaracterisations and flaws in sceptic arguments. However, the relegation of such a large community of sceptics/deniers to the relative darkness of internet message boards may be counter-productive in the long term if broad social coalitions supportive of climate action are to be formed.

## Funding Crisis in Journalism and Its Effect on Climate Change Coverage

Over 100 job losses across the Irish news media sector were announced in May 2019 (Slattery 2019), reflecting wider difficulties in the business model of news (Curran 2019). More recently, state broadcaster RTÉ announced a range of cost-cutting measures, including 200 job losses (Crowley 2019), while Mediahuis, the new owners of Independent News and Media, the country's largest media company, continue to cut costs (Paul 2019). Under-resourced newsrooms tend to rely more on 'information subsidies' (i.e. public relations material) and less on boots-on-the-ground reporting (Picard 2018). In relation to climate change reporting, this means newsrooms and reporters are forced into a reactive role, covering events as they occur rather than planning investigations and in-depth, long-form journalism involving more resource-intensive reporting.

The economic crisis—and to some extent the crisis of legitimacy—in journalism has repercussions for the employment of specialist environment correspondents. The decline in the number of such correspondents has been identified as a factor in declining coverage (Hansen 1994). These specialist reporters play a key role in newsrooms in advising on coverage and policy: they are freer to express opinions, have more autonomy in the newsroom, and are regarded as independent subject experts whose judgement can guide coverage (Schudson 2001; Hiles and Hinnant 2014). Research shows that, during a period in which the *Irish Times* did not have an environment correspondent, climate change coverage declined in frequency, and when the *Irish Independent* appointed a correspondent, its level of coverage increased (Robbins 2019).

The challenges faced by media organisations caused by continual disruption of their business model impact on their ability to produce quality journalism, yet the picture, at least when it comes to environmental coverage, is more nuanced. Newsrooms appear to have acknowledged the increased importance of climate change in the policy and political agenda. For instance, RTÉ initially appointed journalist George Lee to a dual role as 'environment and agriculture' correspondent (RTÉ 2014), a combination many saw as problematic. More recently, these roles have been separated, with Lee now working as 'environment and science' correspondent (Halpin 2019). There had also been fears that the departure of the *Irish Independent*'s correspondent Paul Melia to work as a ministerial advisor would diminish coverage published by Independent News & Media (INM) titles. However, the appointment of former *Irish Examiner* reporter Caroline O'Doherty to this role has ensured that coverage has continued. Furthermore, the appointment of Kevin O'Sullivan as environment editor with the *Irish Times* in 2017 has seen an increase in attention for climate change within that publication.

## Conclusion

Ireland is in several ways typical of many territories with regard to the levels and nature of media coverage of climate change. Irish coverage exhibits similar patterns of increase and decline as elsewhere, and Irish coverage is framed in ways familiar from other territories. However, Ireland's coverage is subject to and constrained by its own peculiar influences. Ireland is a small media territory, affected by competition from publications based in the neighbouring territory; its media companies are likewise modestly sized, lacking the reach and resources of global news brands such as the Guardian Group or the *New York Times*, and therefore less equipped to undertake the deep reporting, data visualisation, and multimedia journalism possible in larger organisations. The funding crisis in Irish journalism has been acute, and the dearth of specialist environment correspondents will continue to affect the extent and quality of climate change coverage.

Counter-balancing these concerns is a new urgency around climate action in public discourse. Evidence suggests that, especially when newsrooms lack the resources to pursue their own news agendas, journalists need 'news hooks' on which to base coverage (Ungar 2014). One of the differences between 1988 (when the US media published extensive coverage of climate change) and 2012 (when media attention was low) was the lack of events, policy announcement, debates, and other inputs from claimsmakers (Ibid.). In Ireland, the deliberations of the Citizens' Assembly, the hearings of the Joint Oireachtas Committee on Climate Action, the protests by schoolchildren around the country, and the protests (and creative media strategies) of the Irish section of the Extinction Rebellion movement have provided such 'news hooks' to an unprecedented extent. Further research is needed to examine the extent of this new wave of coverage and to explore how these new climate change constituencies are attempting to engage a range of publics with the issue.

## Note

1. Schmidt et al. studied coverage over a different timescale to Robbins and recorded coverage in the *Irish Times* only; the *Times* has been identified as publishing more climate coverage than its competitors (Fox and Rau 2016). Schmidt et al. found that Irish coverage was 0.27% of total news coverage in 1997–2000; 0.51% in 2001–2005, and 1.82% in 2005–2009 (Schmidt et al. 2013).

## Bibliography

Antilla, L. 2005. Climate of Scepticism: US Newspaper Coverage of the Science of Climate Change. *Global Environmental Change* 15 (4): 338–352.

Bacon, W., and C. Nash. 2012. Playing the Media Game: The relative (In)visibility of Coal Industry Interests in Media Reporting of Coal as a Climate Change Issue in Australia. *Journalism Studies* 13 (2): 37–41.

Boucher-Hayes, P. 2017. Irish Climate Science Forum. RTÉ. Available at: https://www.mixcloud.com/rtéradioplayerlatestpodcasts/drivetime-irish-climate-science-forum/

Boykoff, M.T., and J.M. Boykoff. 2004. Balance as Bias: Global Warming and the US Prestige Press. *Global Environmental Change* 14 (2): 125–136.

———. 2007. Climate Change and Journalistic Norms: A Case-Study of US Mass-Media Coverage. *Geoforum* 38 (6): 1190–1204.

Boykoff, M., J. Katzung, and A. Nacu-Schmidt. 2019. *Media and Climate Change Observatory Monthly Summary: Scientists fear Arctic heating could trigger a climate 'tipping point'*. Issue 27, March. Media and Climate Change Observatory Monthly Summaries.

Carvalho, A. 2007. Ideological Cultures and Media Discourses on Scientific Knowledge: Re-reading News on Climate Change. *Public Understanding of Science* 16 (2): 223–243.

Carvalho, A., M. van Wessel, and P. Maeseele. 2016. Communication Practices and Political Engagement with Climate Change: A Research Agenda. *Environmental Communication* 11 (1): 122–135.

Chalaby, J. 1996. Journalism as an Anglo-American Invention. A Comparison of the Development of French and Anglo-American Journalism. *European Journal of Communication* 3: 303–326.

Cohen, B. 1963. *The Press and Foreign Policy*. Princeton: Princeton University Press.

Crowley, S. 2019. RTÉ to Cut Jobs, Pay and Some Services to Address Financial Crisis. *RTÉ News website*, November 7. Available at: https://www.rte.ie/news/ireland/2019/1106/1089209-rte-cuts/

Cullinane, M., and C. Watson. 2014. *Irish Public Service Broadcasting and the Climate Change Challenge*. Report for RTÉ Audience Council, Dublin.

Curran, J. 2019. Triple Crisis of Journalism. *Journalism* 20 (1): 190–193.
De Blasio, E., and M. Sorice. 2013. The Framing of Climate Change in Italian Politics and Its Impact on Public Opinion. *International Journal of Media and Cultural Politics* 9 (1): 59–69.
Demeritt, D. 2001. The Construction of Global Warming and the Politics of Science. *Annals of the Association of American Geographers* 91 (2): 307–337.
Devitt, C., and E.O. Neill. 2017. The Framing of Two Major Flood Episodes in the Irish Print News Media: Implications for Societal Adaptation to Living with Flood Risk. *Public Understanding of Science* 26 (7): 872–888.
Dirikx, A., and D. Gelders. 2010. To Frame Is to Explain: A Deductive Frame-Analysis of Dutch and French Climate Change Coverage During the Annual UN Conferences of the Parties. *Public Understanding of Science* 19 (6): 732–742.
Djerf-Pierre, M. 2012. The Crowding-Out Effect: Issue Dynamics and Attention to Environmental Issues in Television News Reporting Over 30 Years. *Journalism Studies* 13 (4): 499–516.
Duan, R., and B. Takahashi. 2017. The Two-Way Flow of News: A Comparative Study of American and Chinese Newspaper Coverage of Beijing's Air Pollution. *International Communication Gazette* 79 (1): 83–107.
Feldpausch-Parker, A.M., et al. 2015. Spreading the News on Carbon Capture and Storage: A State-Level Comparison of US Media. *Environmental Communication* 7 (3): 336–354.
Fox, E., and H. Rau. 2016. Climate Change Communication in Ireland. *Oxford Research Encyclopedia of Climate Science*: 1–32.
Gibbons, J. 2017. *New Climate Science Denial Group Launches in Ireland, Desmog UK*. Available at: https://www.desmog.uk/2017/05/05/new-climate-science-denial-group-launches-ireland. Accessed 12 Sept 2017.
Goffman, E. 1974. *Frame Analysis: An Essay on the Organization of Experience*. New York: Harper & Rowe.
Government of Ireland. 2017. *National Mitigation Plan*. Dublin.
Government of New Zealand. 2019. *New Zealand's Greenhouse Gas Inventory 1990–2017*.
Hallin, D.C., and P. Mancini. 2004. *Comparing Media Systems: Three Models of Media and Politics*. Cambridge: Cambridge University Press.
Halpin, H. 2019. Fran McNulty to Replace George Lee as RTÉ's Agriculture and Consumer Affairs Correspondent. *The Journal.ie*, June 11.
Hansen, A. 1994. Journalistic Practices and Science Reporting in the British Press. *Public Understanding of Science* 3 (2): 111–134.
Hiles, S.S., and A. Hinnant. 2014. Climate Change in the Newsroom: Journalists' Evolving Standards of Objectivity When Covering Global Warming. *Science Communication* 36 (4): 428–453.

Hirji, Z. 2017. Climate Contrarian Gets Fact-Checked by MIT Colleagues in Open Letter to Trump. *Inside Climate News*, March 6. Available at: https://insideclimatenews.org/news/06032017/climate-change-denial-scientists-richard-lindzen-mit-donald-trump. Accessed 12 Sept 2017.

Jaspal, R., and B. Nerlich. 2014. Fracking in the UK Press: Threat Dynamics in an Unfolding Debate. *Public Understanding of Science* 23 (3): 348–363.

Kenny, E. 2015. National Statement Delivered by An Taoiseach Enda Kenny to the 21st Conference of the Parties to the UN Framework Convention on Climate Change. Available at: http://unfccc.int/files/meetings/paris_nov_2015/application/pdf/cop21cmp11_leaders_event_ireland.pdf

Kirilenko, A.P., T. Molodtsova, and S.O. Stepchenkova. 2015. People as Sensors: Mass Media and Local Temperature Influence Climate Change Discussion on Twitter. *Global Environmental Change* 30: 92–100.

Lyytimäki, J. 2011. Mainstreaming Climate Policy: The Role of Media Coverage in Finland. *Mitigation and Adaptation Strategies for Global Change* 16: 649–661.

McCombs, M.E. 2004. *Setting the Agenda: the Mass Media and Public Opinion*. Cambridge: Polity Press.

McKewon, E. 2012. Talking Points Ammo: The Use of Neoliberal Think Tank Fantasy Themes to Delegitimise Scientific Knowledge of Climate Change in Australian Newspapers. *Journalism Studies* 13 (2): 277–297.

McNally, B. 2015. Media and Carbon Literacy: Shaping Opportunities for Cognitive Engagement with Low Carbon Transition in Irish Media, 2000–2013. *Razon y Palabra* 91: 2000–2013.

Mercille, J. 2014a. The Role of the Media in Fiscal Consolidation Programmes: The Case of Ireland. *Cambridge Journal of Economics* 38 (2): 281–300.

———. 2014b. The Role of the Media in Sustaining Ireland's Housing Bubble. *New Political Economy* 19 (2): 282–301.

Mouffe, C. 2005. *On the Political: Thinking in Action*. London/New York: Routledge.

Nelkin, D. 1987. Selling Science: How the Press Covers Science and Technology. *Physics Today* 43: 41–49.

Newman, T.P. 2017. Tracking the Release of IPCC AR5 on Twitter: Users, Comments, and Sources Following the Release of the Working Group I Summary for Policymakers. *Public Understanding of Science* 26 (7): 815–825.

Nisbet, M.C. 2009. Communicating Climate Change: Why Frames Matter for Public Engagement. *Environment: Science and Policy for Sustainable Development* 51 (2): 12–23.

O'Brien, C. 2019. YouTubeTerminates Gemma O'Doherty's Account Over Breach of "Hate Speech" Policy. *Irish Times*, July 16.

O'Neill, S., H.T.P. Williams, et al. 2015a. Dominant Frames in Legacy and Social Media Coverage of the IPCC Fifth Assessment Report. *Nature Climate Change* 5 (4): 380–385.

O'Neill, S., H. Williams, et al. 2015b. Dominant Frames in Legacy and Social Media Coverage of the IPCC Fifth Assessment Report – Supplementary Material. *Nature Climate Change* 5 (2): 1–9.

Painter, J., and T. Ashe. 2012. Cross-National Comparison of the Presence of Climate Scepticism in the Print Media in Six Countries, 2007–10. *Environmental Research Letters* 7 (4): 044005.

Painter, J., and N.T. Gavin. 2015. Climate Skepticism in British Newspapers, 2007–2011. *Environmental Communication* 10 (4): 37–41.

Paul, M. 2019. The More It Changes at INM, the More It Stays the Same. *Irish Times*, October 25.

Pearce, W., et al. 2014. Climate Change on Twitter: Topics, Communities and Conversations about the 2013 IPCC Working Group 1 Report. *PLoS One* 9 (4): 1–11.

———. 2019. The Social Media Life of Climate Change: Platforms, Publics, and Future Imaginaries. *Wiley Interdisciplinary Reviews: Climate Change* 10 (2): 1–13.

Picard, R.G. 2018. The Economics of Journalism and News Provision. In *Journalism*, ed. T.P. Vos, 281–297. Berlin: Walter de Gruyter.

Rahmstorf, S. 2004. The Climate Sceptics. *Potsdam Institute for Climate Impact Research*, pp. 76–82. Available at: http://www.pik-potsdam.de/stefan/Publications/Other/rahmstorf_climate_sceptics_2004.pdf

Readfearn, G. 2017. Trump's Potential Science Adviser William Happer: Hanging Around with Conspiracy Theorists. *The Guardian*, February 21.

Robbins, D. 2016. Morality Versus Politics: The Irish Media and Laudato Si. *Studies: An Irish Quarterly Review* 105 (420): 441–450.

———. 2019. *Climate Change, Politics and the Press in Ireland*. London: Routledge.

RTÉ. 2014. RTÉ News Announces George Lee as New Agriculture and Environment Correspondent. *RTÉ Press Centre*, March 13. Available at: https://presspack.rte.ie/2014/03/13/rte-news-announces-george-lee-as-new-agriculture-and-environment-correspondent/. Accessed 8 Sept 17.

Schäfer, M.S., and I. Schlichting. 2014. Media Representations of Climate Change: A Meta-Analysis of the Research Field. *Environmental Communication: A Journal of Nature and Culture* 8 (2): 142–160.

Schmidt, A., A. Ivanova, and M.S. Schäfer. 2013. Media Attention for Climate Change Around the World: A Comparative Analysis of Newspaper Coverage in 27 Countries. *Global Environmental Change* 23 (5): 1233–1248.

Schudson, M. 2001. The Objectivity Norm in American Journalism. *Journalism* 2 (2): 149–170.

Shabecoff, P. 1988. Global Warming Has Begun, Expert Tells Senate. *The New York Times*, June 24.

Slattery, L. 2019. Recent Job Losses Reflect Difficult Outlook for Irish Journalism. *Irish Times*, May 28. Available at: https://www.irishtimes.com/business/media-and-marketing/recent-job-losses-reflect-difficult-outlook-for-irish-journalism-1.3906066

Smith, M. 2019. Village Editorial, September: Gemma O'Doherty 2019. *Village Magazine*, September 11.

Ungar, S. 2000. Knowledge, Ignorance and the Popular Culture: Climate Change Versus the Ozone Hole. *Public Understanding of Science* 9: 297–312.

———. 2014. Media Context and Reporting Opportunities on Climate Change: 2012 Versus 1988. *Environmental Communication: A Journal of Nature and Culture* 8 (2): 233–248.

Veltri, G.A., and D. Atanasova. 2017. Climate Change on Twitter: Content, Media Ecology and Information Sharing Behaviour. *Public Understanding of Science* 26 (6): 721–737.

Wagner, P., and D. Payne. 2015. Trends, Frames and Discourse Networks : Analysing the Coverage of Climate Change in Irish Newspapers. *Irish Journal of Sociology* 2: 1–24.

Williams, H.T.P., et al. 2015. Network Analysis Reveals Open Forums and Echo Chambers in Social Media Discussions of Climate Change. *Global Environmental Change* 32: 126–138.

Wilson, K.M. 1995. Mass Media as Sources of Global Warming Knowledge. *Mass Communications Review* 22 (1 & 2): 75–89.

# Cultural and Visual Responses to Climate Change: Ecological Reading of Irish Zombie Movies

*Pat Brereton*

## Introduction

Eco-film criticism's over-arching purpose should not be to impose a political programme but to help create public spaces for debate and argument over the claims of the environment for a place in political life (see Rust et al. 2013). This proposition certainly rings true from an Irish perspective, where we have a relatively small audio-visual media catalogue to explore. Fictional narratives, including a recent growth in low-budget horror flicks like *Dead Meat*, can call attention to the under-belly of agriculture and the food industry in a country where this is a major cause of concern with increasing levels of methane and $CO_2$ emissions. Such low-budget horror-zombie films use landscape in a less celebratory or nostalgic mode and actively call attention to the growth of environmental risk and

P. Brereton (✉)
School of Communications, Dublin City University, Dublin, Ireland
e-mail: pat.brereton@dcu.ie



D. Robbins et al. (eds.), *Ireland and the Climate Crisis*, Palgrave Studies in Media and Environmental Communication,
https://doi.org/10.1007/978-3-030-47587-1_11

food security, which casts a long shadow over contemporary factory farming.

In particular, these narratives address new audiences, especially those who might not be predisposed to recognising the long-term seriousness of climate change. These readings can further assist in the evolving discipline of eco-film studies and hopefully bolster more active engagement with a broader range of audiences, while calling attention to local issues, using the creative imaginary of horror.

The evolution of ecocinema can be traced back to Scott MacDonald, who argues that only certain experimental films can promote an ecocentric sensibility and considers the 'fundamental job of ecocinema as a retraining of perception, as a way of offering an alternative to conventional media-spectatorship' (cited in Rust et al. 2013: 45). Paula Willoquet-Maricondi (2010) similarly suggests that only certain types of independent lyrical and activist documentaries may be thought of as ecocinema, because they are the most capable of inspiring progressive eco-political discourses and action among viewers. At one level, of course, both scholars are correct. Nonetheless, it is more beneficial, according to a number of environmentally focused film volumes (including Rust et al. 2013; Weik von Mossner 2017) to broadly accept that *all* types of film, from the excesses of Hollywood blockbusters to the most rarefied and explicitly ecological art-house cinema, assist in their own way towards foregrounding ecological issues and concerns.

Ireland affords a useful test site for environmental media, especially taking into account its poor record in facing up to climate change. This is especially evident within agriculture in particular, as a primal rural profession, which also serves as a barometer of ever-changing ecological land ethics. Historically, farmers have displayed a deep sense of place and environmental stewardship, as evident in seminal Irish classics like *Man of Aran* (1934), *The Quiet Man* (1952) and *The Field* (1990), while contemporary low-budget zombie and horror fantasy films like *Isolation* (2005), *Dead Meat* (2005), *Boy Eats Girl* (2005), *Wakewood* (2009) and *Grabbers* (2012) all tend to subvert and disrupt this consensual view. Such B-rated films have had little if any scholarly analysis of their thematic and aesthetic strategies in subverting a general celebration of nature and farming, while alternatively highlighting environmental concerns around food production and consumption.

This chapter examines eco-textual readings of Irish fictional films which both overtly and covertly address local climate change concerns. It particularly situates representations of place, coupled with cautionary tales around food production and farming, through contemporary low-budget Irish horror movies.

At the outset, more conventional economic and scientific modes of thinking tend to privilege, if rarely cautioning against, a techno-fix approach to environmental problems, from food security to the growth of $CO_2$ emissions and methane in particular. These techno-fix tensions are often foregrounded and situated through the use of genetically modified organisms (GMO), as seen through the representation of cows in contemporary horror tales like *Isolation*. *Dead Meat*, most notably, calls attention to pervasive worries about unsustainable agricultural growth and the ever-present spectre of climate change, where necessary checks and balances around production and distribution are ignored. Instead, blunt economic growth remains the dominant principle and goal for all forms of development, a fixation which leads to dangerous consequences that have been obfuscated, if not ignored, across more mainstream representations of agriculture on film.

Over recent years, environmental scholars have begun to tease out the power of mass media and film in foregrounding and privileging a broad range of environmental discussion and debate like those cited above. Such studies focus especially on liminal and troubled landscapes, rather than simply concentrating on the sublime pleasures of wild rural topographical features of the landscape. One might position horror and its zombie film offshoots as a useful sub-genre that helps to highlight the dangers inherent in excessive production in agriculture, while providing a more critical engagement with romantic representations of landscape especially. This can be read as part of a critical environmental discourse and consequently help to puncture the uncontested celebration of farming stewardship of the land.[1]

To call attention to such dramatic paradoxes in a creative and engaging manner, horror allegories of zombies, alongside genetic engineered (GMO) farm animals, all serve to dramatise these very prescient environmental fears around food security and sustainable, even organic, farming. The B-rated storylines work as cautionary tales that visually address such a potentially shocking future scenario which might be activated if society does not urgently face up to the realities of environmental food production and the harvesting of other natural resources from a sustainability perspective.[2]

Yet, from a 'business-as-usual' and industrial perspective, farming and agriculture constitutes a dominant sector of the Irish economy and the one which is most resistant to change—a change which is necessary to achieve a low-carbon future.[3] In response, deep-seated, if probably unconscious fears of despoiling nature are actively registered across several Irish dystopic horror-zombie films. For example, *Shrooms* (2007) focuses on a group of young, hippie-like American tourists who ostensibly espouse environmental values and who apparently wish to 'be at one with nature'. But the storyline turns the conventional notion of a tourist eco-trip on its head when the protagonists inhale wild mushrooms picked from the 'magical' landscape of Glengariff Forest Park. Consequently, the narrative's evocation of an erstwhile romantic landscape is radically subverted—recalling generic Hollywood classics like *Deliverance* (1972) and its malevolent evocation of primeval natives from middle America, or the very different contemporary European take on environmental cults in *Midsommar* (2019). At one stage, early in the trip and feeding into the most regressive post-colonial stereotypes, two natives in dishevelled dirty clothes appear almost grunting, while trying to communicate with the visitors, bearing hatchets and seeking to pick up the road-kill left for food. Looking on in shock, the visitors talk of the place being like *The Island of Doctor Moreau* (1996), using well-understood generic short-hand and intertextual signifiers to describe such a primitive and horrific habitat.

Ireland's current agricultural strategy, encapsulated in Food Harvest 2020, commits to constantly growing the value of dairying and the beef sector, recalling for some the excesses of so-called zombie banks and other institutions that facilitated a major financial crash in Ireland almost a decade ago. Most notably, this strategy directly contradicts government's environmental policy, designed to meet European regulations and to radically reduce carbon emissions—including methane from cattle—following the 2015 Paris Agreement. This industrial-level uncritical strategy of exponential growth in agriculture, designed to simply maximise the economic productivity of the land in spite of the environmental consequences, remains a long way from the small-scale productive nurturing of soil and farmland explored in *The Field* and dramatised most effectively in the very realistic exposition around the insecurity of farming in *Pilgrim Hill* (2012),[4] much less recognising the organic sourcing of wild mushrooms in *Shrooms* or even recalling the worried farmer striving to survive in *Isolation*, who reluctantly agrees to illicit medical experimentation on his cows for financial gain. Certainly, all of this unregulated development is a long way away from a deep environmentalist's embrace of sustainable green growth,[5] much less a radical 'no growth' or 'steady-state' economy, as exemplified by Tim Jackson's ***Prosperity without Growth*** (2009).

When it comes to the environmental agenda, new modes of media representations in particular have reflected and shaped the contested existing visions around the human-nature relationship (Parham 2016), creating engaging narratives and generic formats for new generational audiences. Of course, this remains a key focus for an examination of such zombie movies. Certainly, far greater understandings of the power of such fictional stories are needed through further research and scholarship, coupled with an empirical investigation of how new audiences consume such media especially from an environmental perspective (see Brereton and Gomez 2020; Weik von Mossner 2017).

## The Irish Turn to Environmentalism: From Classic Literature to Zombies

A study by Robert Brazeau and Derek Gladwin (2014) seeks to decode the classic writings of James Joyce as part of a critical second (or even third) wave of environmental literary criticism and pro-active engagement, while alternatively suggesting that William Butler Yeats and other contemporary poetic writers—to which I could include much mainstream Irish cinema—remain locked into a first-wave register of romantic and biophilic engagement, with the foregrounding love of (wild) nature. This first wave of eco-criticism sometimes unconsciously strives to support the necessity of moving to a low-carbon society, while remaining centrally preoccupied with conserving a wild uncontaminated landscape and habitat. Consequently, over the years, environmental scholars have drawn on pervasive and long-term narrative tropes and discourses that privilege romantic sensibilities, especially through recalling rewilding projects, coupled with a biophilic love of nature. But according to some, this approach steers a path towards essentialising and even reifying nature.[6]

These environmental tensions are strongly inferred and referenced across many Irish horror tales, suggesting that there is a lot to lose in accepting ongoing demands to transform and modernise the landscape, which over the centuries has been alternatively ravaged by British colonisation and subsequently through more insidious forms of global capitalism. Consequently, if only unconsciously, such alternative low-budget horror-zombie storylines can be recognised as striving to re-imagine a future that supports sustainable low-carbon energy, alongside valorising varying alternative models of organic food production.

## Environmental Zombie Movies, Survivalism and Eco-textual Analysis of Food Production

It is often said that the apocalypse has been embedded in our collective consciousness, for it functions as myth and a narrative device used across popular culture while performing many functions within the contemporary world. In her essay on the Irish horror film, Radley (2013: 117) argues that the genre calls on tropes from the apocalypse which has helped transform Irish cinema. While probably overstating the format's radical nature, Radley suggests that 'the dialectic that is established in Irish horror films between a disruptive, semiotic generic sensibility and an established, symbolic national sensibility works to deconstruct and rearticulate the post-colonial bias in the discourse of Irish national cinema'. At least, Radley is accurate in positing that these films do reflect a 'crisis in/of the representable in the wake of the more general transformation in the landscape of Irish subjectivity post-Celtic Tiger' (2013: 123).

Negra further believes that these films 'radically disrupt other representational conventions, notably of Ireland as a recuperative space' (see Negra 2007: 143); all of which is a long way from more conventional mainstream farming and its romantic representations discussed above. The journey to Ireland, or more easily from the city to the countryside, 'no longer facilitates psychic healing (as it does in the tourist films) or the return to a safe past, but instead becomes a voyage into terror and death' (Barton 2019: 85).

Essentially, using the lens of the commercial exploitation of environments around food production and disturbing the long-held belief of farming stewardship within mainstream cinema, such tensions can help to make environmental issues more resonant for mass audiences. Yet, while there is always a danger of such concerns being brought to the fore simply to become populist, one could counter-argue that some exemplary Irish zombie films nonetheless can foreground the environmental agenda in a more direct and corporeal manner, especially for non-committed and ostensibly non-green audiences, all the while dramatising these prescient environmental concerns and tensions through a powerful aesthetic and by further using a memorable allegorical format.

As a genre, zombie movies are grounded in themes of collapse, survival, security and resource scarcity which help ensure commercial media-rendered tales of the undead can effectively present environmental issues. Drezner (2011), most notably, teases out contrasts between zombies, individual and collective human behaviour, while others highlight how, at least since George Romero's *Night of the Living Dead* (1968), the genre has remained preoccupied with dramatising 'society's subaltern and out-casts', most frequently displayed in anonymous shopping centres. Furthermore, their cannibalism can be explained as encapsulating the need for the undead to consume the living as 'payback for the pains they suffered in life' (cited in Murphy 2017: 9). From the view of the zombie's *prime directive*, the notion of 'nature's revenge' for some environmental disasters seems particularly relevant and potent in this current climate emergency. These all-encompassing tensions can be seen as building on James Lovelock's (2009) thesis that the Earth is a living, dynamic inter-connected system, which he termed 'Gaia', while noting how it does not passively suffer oppositions from organisms that do it harm. Meanwhile, zombies apparently do not seem to have this blind spot. Hence, the evolving zombie sensibility might paradoxically imply a new and more fruitful even healthy relationship with the earth.

Other scholars have provided insight into how such sensational 'image events', like the rise of the zombie apocalypse and the horror movie generally, can serve as a form of transformative environmental politics (DeLuca 2006). Patrick Murphy (2017), in turn, posits a range of possible environmental visions being developed through recurring tropes (e.g., 'wilderness', 'pollution'), narratives (such as 'jeremiad and apocalypse') and metaphors ('a place apart, tipping point, mother earth and so on'), all of which are evident in art, literature and media (Corbett 2006; Heise 2008).[7] Zombie and horror movies can most pointedly provide a clearly defined and a useful new popular cultural prism to help tease out some of these shocking but broad-ranging environmental tensions.

## Dead Meat: Irish Zombies and the Celtic Tiger Economy!

The environmental food agenda of weaning publics off meat remains a major challenge for the affluent world and especially across food-producing countries like Ireland. Anchored to debates around genetic modification, metaphors like 'Frankenfoods' and commonplace notions like 'playing God' serve to close off debate around the potential applications of synthetic meat. Whereas asking factual and rhetorical questions can potentially help weigh up 'its risks and benefits, and envisage changing current mentalities or behaviour in order to adapt to scientific developments, enabling a consideration of synthetic meat's possible implications for agriculture, environment and society' (Marcu et al. 2015: 1).

Sean Crosson (2019), in his study of Gaelic sports, provides an insightful cultural reading of the film's use of hurling sticks and a rejection of previous touristic depictions of Ireland. Remaining heavily indebted to previous international horror films, *Dead Meat* like others mentioned above, depicts rural Ireland as a 'place far removed from the welcoming and reassuring locale represented in tourist iconography and contemporary Irish postcards' (166). A Spanish tourist, Helena (Marian Araujo), is lost in the Leitrim countryside and assisted by the local gravedigger, Desmond (David Muyllaert), and trainer of the local underage hurling team Cathal Ceaunt (Eoin Whelan). Helena survives her ordeal, however, only to be locked up in a truck, like a farm animal, at the film's close (Crosson 2019: 166).

Crosson explores how the iconic hurley stick, which is used in traditional Gaelic Athletic Games (GAA), can be decoded as 'the new chainsaw'—beloved by the horror gore genre—while at the same time recalling the mythological Irish figure of Cuchulainn and his heroic exploits. Most interestingly, the critic draws comparisons between the figure of the zombie as a highly suggestive motif of the Celtic Tiger, which encapsulated a period of excessive economic growth in the country, including a rise in intensive monocultural farming practices, all of which is certainly not compatible with sustainable environmental practices. Hence, the explanation offered for the infection of humans and their transformation into zombies is rationalised through the spread of a particularly virulent strain of Bovine spongiform encephalopathy (BSE) or mad cow disease, from animals to humans, following the feeding of livestock with food acquired from the carcasses of dead animals. Director Conor McMahon is referring here to the practice identified by scientists as responsible for the actual spread of BSE, while also providing a wider critique of an unsustainable means of food production. As Felicity Lawrence notes in her 2008 Penguin study *Eat Your Heart Out: why the food business is bad for the planet and your health*:

> Modern food production involves processes that, quite apart from having little care for real nutrition, drive people off the land, stimulate migration, increase inequalities and the depth of poverty are corrosive of society and depend on extravagant use of natural resources, from water to oil to land, that are running out. The politics of food is in other words not the art of shopping but the politics of modern globalised capitalism itself. (quoted in Cockburn 2008)

Like many zombie movies, *Dead Meat* dramatically calls attention to a visceral critique of capitalism itself, while explicitly speaking to the so-called Celtic Tiger period, which was marked by valorising ostentatious wealth and high consumerist spending, encouraged by the neoliberal economic policies of successive Fianna Fáil-led governments. This was followed by a serious recession and resulted, according to Central Bank reports, in the country having one of the highest levels of both national and personal debt in Europe. The director could hardly have chosen a more appropriate moment to employ such a resonant motif. The protagonist Cathal Cheunt's defence of the few uninfected survivors from an attacking mad cow with his hurling stick and sliotar, in the style of Cuchulainn, certainly stands out. While retrieving the sliotar, Cheunt utters—'a lot done, more to do'. These were the precise words of a political campaign slogan used by Fianna Fáil and subsequently became associated with the party's then leader Bertie Ahern, prior to the 2002 election which returned the party to power. McMahon quips 'that's a line only Irish people will probably get' (Crosson 2019: 171).

All the while, consumer worries about unsustainable agri-food technologies, including genetically modified organisms (GMOs), cloning and nanotechnology, have continued at pace, while generally shown to be underpinned by considerations of unnaturalness, a lack of trust in science, risk management provision, ethics, uncertainty and unknown long-term effects, alongside concerns about the wider implications of science for society (Siegrist 2008). Certainly, concerns about what is ***natural*** underpin many mainstream Irish audio-visual narratives, not to mention the growing dilemma around climate change. Arguments around 'interfering with nature' are seldom foregrounded, except in such cautionary tales, and this response also is common in representations of biotechnology including GMOs (Tenbult et al. 2005) or cloning (Shepherd et al. 2007), and possibly underpins the rejection of food technologies such as cloned beef (Marcu 2015: 549).

## Environmental Implications

Much is clearly out of kilter between humans and nature, as evident by the planet's ongoing difficulties around climate change (see chapter "Climate Change in Ireland: Science, Impacts and Adaptation" by Sweeney, this volume), and the creative audio-visual imaginary has a key role to play in providing new and more powerful ways of representing such fears and difficulties. Facing up to global apocalypse, and taking into account the traumatic implications for nature (human and environment), remains the most pervasive generic aesthetic and environmental political agenda. In its own small way, *Dead Meat* and other zombie films can assist in calling attention to such deep-seated environmental fissures in various ways.

At the same time, however, critical eco-scholarship must try to constantly gauge if the media commons is simply being complicit in promoting a 'business-as-usual' mode of greenwashing, or indulging in a romanticisation of rural environmental values, as witnessed across much mainstream Irish cinema or, alternatively, maybe affording some chink of light in this otherwise polarised and monocultural representation of farming. Following in the footsteps of Stuart Hall and other cultural theorists, environmental communication scholars can uncover some positive, yet often contradictory media discourses across popular film culture to help address these complex environmental tensions. Of course, this mode of eco-textual analysis remains somewhat tentative and open to criticism, while ostensibly motivated towards promoting a more environmentally sustainable kind of society and even actively inspiring new models of human agency.[8] Constructing a possible template for effective environmental engagement using fictional narratives remains the primary objective of such scholarship, while always being wary of the perennial dangers of romanticising and greenwashing, as highlighted by Toby Miller (2017) among others.

Environmental film scholars actively seek to discover how our media landscapes are being shaped. According to Cox (2010: 51), this is a necessary part of the 'challenge of building a more sustainable world in the face of disruptive or unsustainable social and economic systems'. Nevertheless, this is a leap of imagination for environmental media scholarship generally, while even contemplating the potential uses of valorising such small-scale and B-rated horror-science fiction and zombie narratives is a big task. Within the broad spectrum of environmental communications and general literacy, in placing climate change at the top of the political social and cultural agenda, urgent and concerted action is badly needed to mobilise a variety of responses for change. A broad range of generic media formats, including horror-zombie movies, can be usefully co-opted in this ongoing struggle to develop a robust curriculum for environmental media studies and help support transformational change (see Brereton 2019).

There certainly is no magic bullet audio-visual stimulus available to speak to all types of people/citizens, and most certainly, a full range of generic protocols and narrative tropes need to be constantly co-opted and re-purposed in this endeavour. As this review chapter illustrates, various forms of apocalyptic and horror fiction can almost counter-intuitively help to both privilege and contest an environmental and ethical evocation of place and a deep celebration of habitats, including an uncontested level of stewardship of the land which farmers have enjoyed. At the same time, such commercially conceived horror flicks, which are designed for the teenage youth market, can further serve a useful role in helping to puncture the romanticisation of land, farming and especially factory food production. Without simply preaching at farmers, or the tourist industry for that matter, and in calling out political agents for their lack of environmental rectitude, such visceral and generic narratives can alternatively help communicate a potent environmental message around the urgent need for behavioural transformation. Calling on the intrinsic power of the media apparatus to help trigger and shape audience perspectives towards a much-needed tipping point, such diverse storylines can at least assist in shining a light on this prescient but very complex environmental struggle.

As mentioned in *Being Ecological* and recalling the current concern over plastics in the oceans, Timothy Morton (2018: 77) imagines 'all the plastic bags in existence' and 'all that will ever exist, everywhere. This heap of plastic bags is a hyperobject', which remains interconnected with us across time and space. Similarly, potent audio-visual generic narratives, including the corporeal gut-wrenching genre of horror-zombie movies, in their own unique way can assist in encouraging even non-committed audiences towards recognising and hopefully in the future actively engaging with such evocative environmental food issues.

One could almost hypothesise that a major cautionary exercise embedded within such provocative horror storylines regarding food production is focused on the move towards making agricultural and sustainable production/consumption more appealing to citizens, without degenerating into preaching, finger-wagging, moralising and fear. While many scholars believe Hollywood de facto embodies and displays a reductive level of engagement with regard to environmental exposition (Cubitt 2017; Miller 2017; Moore 2017), there remains some hope, in spite of dominant neoliberal commercial pressures, that generic and mainstream film can, in their own unique way, help to sow the seeds of a progressive environmental discourse.

Nevertheless, recent media research (Garforth 2018; Hoggett 2019) has demonstrated that simply inducing fears of an apocalypse, much less appealing to the zombie apocalypse in this case, does not necessarily work as expected on audiences and alternatively can quickly lead to worry, fatalism or even compassion fatigue. This remains an ongoing concern for mainstream eco-film scholarship, alongside political activism like Extinction Rebellion, while extolling the apparent benefits of representing spectacular and/or dystopic or horror-based textual imaginaries in communicating science and its environmental fault lines. While recognising relatively poor levels of political engagement by mass audiences with climate change issues, one could suggest this demands a range of pro-active narrative strategies and fictional protocols that more actively speak to what might be typified as more passive and mainstream viewer sensibilities. Such audiences might not necessarily be predisposed towards supporting environmental rectitude per se, much less embracing radical behavioural change for that matter, but hopefully the seeds of environmental disquiet and learning are sown through such disruptive storylines.

## Conclusion

At a geographical and also at a symbolic level, one hopes that the small island of Ireland can be centrally involved in future debates and research around (European) environmentalism, sustainability and representations of landscape, while also addressing more contentious problems of food production and security, as well as supporting low-carbon energy production. The filmmaker as 'surrogate witness for the filmgoers' (Strain 1998:

162) can capture this political and environmental transition and moment of sublime synthesis in an image that conforms with audience's pre-existing conceptions of the 'real world'.[9] Many historical Irish media discussions, at least from a romantic and a landscape perspective, can be seen to move from a national to a more global environmental sensibility, while instilling powerful images of Irish landscape that in turn help to frame and incite audience's appreciation of environmental and ecological agendas.

Dominant environmental narratives constantly project images of looming catastrophe and collapse. But that 'apocalyptic horizon' (Dryzek 1997: 37) also opens up possibilities for imagining alternatives (Garforth 2018: 18). Alongside news and documentaries, fictional stories of all types—including zombie-horror movies—can help to sow the seeds for 'radical change; the necessity for change; about change that might happen despite our (worst and complicit anti-environmentalist) selves' (Canavan and Robinson 2014: 13). At the same time, we must be constantly aware of the danger, as alluded to earlier, with scholars like Hume (2009: 232–3) and many other behavioural environmental scholars believing that images of 'catastrophe and chaos' are more likely to simply scare citizens into panicked responses. Nonetheless, emotionally resonant (re)presentation in horror and zombie narratives that visualise and call attention to the causes and consequences of climate change—be they global security concerns, genetic fears or more local fears around sustainable agriculture and the preservation of rural communities in Ireland—can at least help to keep the polarising tensions between immersion and displacement in play and, through pleasure and engagement, hopefully help keep the real horrors of climate catastrophe constantly on the screen.

New and robust forms of audience studies and action research are needed to fully understand the markers and trigger-points to effect change across such a universal environmental crisis. This would in turn help to frame appropriate communication strategies that can bring global citizens and more local national groups to appreciate the urgency of facing up to the environmental crisis, while at the same time taking on board the long-term demand for sustainable development into the future.[10] However, such an empirical audience research approach requires dedicated effort and resources that need to be developed over time (see Brereton and Gomez 2020). This introductory chapter simply sets out to record some of these concerns that actively call attention to various interconnecting issues around sustainable food production and security into the future.

## Notes

1. Such micro-horror narratives further foreground environmental risk and the 'Tragedy of the Commons'; recalling concerns for planetary limits, rewilding and, even more controversially, addressing the need to adopt the values of de-growth as a radically new economic norm. But we do not have the space to address these concerns adequately in this short overview.
2. Incidentally, on a much larger canvass, Hollywood science fictional tales of food scarcity and limits sparking planetary disaster have also become a topic of late. Re-imagining, for instance, the contemporary role of farming in producing a potato crop on a desolate alien habitat in *The Martian* (2015), or imagining new modes of aqua-farming in *Blade Runner 2049* (2017), alongside coping with failing crops back on earth in *Interstellar* (2014), while constantly attempting to discover (alien) fertile landscapes, all have become an environmental preoccupation across Hollywood science fictional narratives. Alternatively, a more dystopic cautionary global format like horror and zombie films can help to speak to this urgent need to address upcoming catastrophic environmental changes to our (agricultural) habitats.
3. Furthermore, in protecting the growing tourist industry, which ostensibly promotes a rustic organic mode of agriculture as remaining timeless—and not at the same time embracing images of solar or windfarms dotted across the land and off-shore, which appear to be totally out of kilter with the normal order and traditional romantic beauty of the island's pristine natural landscapes.
4. See eco-reading of the film in Shanagher and Brereton 2019.
5. As Hickel and Kallis conclude their study 'Is Green Growth Possible?', 'it seems likely that the insistence on green growth is politically motivated. The assumption is that it is not politically acceptable to question economic growth and that no nation would voluntary limit growth in the name of the climate or environment; therefore green growth must be true, since the alternative is disaster' (2019: 15).
6. As Lisa Garforth affirms, the 'problem' with the so-called first-wave environmentalism revolves around how 'it should not offer the beauty of landscape or harmony with nature but instead unsettle us by gesturing at the enormous scale of an interconnected universe and the uncanny experience of living without stable ontological categories' (Garforth 2018: 146).
7. 'Historians, geographers and anthropologists have likewise produced scholarship on the capacity of environmental representations to shape public consciousness, tied in significant ways to place, performance and gender in relation to the production and maintenance of cultural identities' (Murphy 2017: 6).

8. For instance, in the 2016 book *Half-Earth: Our Planet's Fight for Life*, Pulitzer Prize-winner Edward O. Wilson makes a case for Green Radicalism, positing that we approach the planet's ills through a strategy modelled on the idea of an ark rather than a lifeboat. See also reading of apocalypse in *Noah* and other related films reviewed in Brereton (2016).
9. 'Viewing the landscape as picturesque or sublime serves to personify the land. In other words, although a filmmaker's or tourist's perception is required to make the vision come to life, the land with its 'authenticity of effect' becomes a narrative character affecting spectator and diegetic character alike. The spectacular landscape may have a pleasurable effect on the spectator at the same time that the land's mysterious effects are foregrounded by the narrative itself' (Strain in Degli-Esposti 1998: 163).
10. Some theorists and most recently activists from the environmental action group 'Extinction Rebellion' talk of a 'tipping point' when new ideas/ attitudes and politics achieve orthodoxy. This phenomenon has been studied with regard to the growth of feminism and anti-racism in particular. For example, it is suggested that once Native Americans reached 10% of total population, it was 'inevitable' that such agendas would become mainstream. With regard to ecology, it is suggested that the 'broad church' of pressure groups need to cohere under a single easily understood agenda, if the green message together with an global attitudinal change is to become acceptable for mass audiences.

## Bibliography


Barton, R. 2019. *Irish Cinema in the Twenty-first Century*. Manchester: Manchester University Press.

Brazeau, R., and D. Gladwin, eds. 2014. *Eco-Joyce: The Environmental Imagination of James Joyce*. Cork: Cork University Press.

Brereton, P. 2005. *Hollywood Utopia: Ecology in Contemporary American Cinema*. Bristol: Intellect Press.

———. 2013. Irish Eco-cinema, A Survey. In *Culture and Media: Ecocritical Explorations*, ed. R.K. Alex. Newcastle Upon Tyne: Cambridge Scholars.

———. 2016. *Environmental Ethics and Film*. Abingdon: Earthscan Routledge.

Brereton, Pat. 2019. *Environmental Literacy and New Digital Audiences*. Abingdon/New York: Routledge.

Brereton, P., and V. Gomez. 2020. Media Students, Climate Change, and YouTube Celebrities: Readings of Dear Future Generations: Sorry Video Clip. *ISLE: Interdisciplinary Studies in Literature and Environment* 27 (2): 385–405, Spring 2020. https://doi.org/10.1093/isle/isaa021.

Canavan, G., and K.S. Robinson, eds. 2014. *Green Planets: Ecology and Science Fiction*. Middletown: Wesleyan University Press.

Cockburn, A. 2008. Review: A Bitter Harvest. *The Sunday Times: Culture*, July 13.

Corbett, J.B. 2006. *Communicating Nature: How We Create and Understand Environmental Messages*. Washington, DC: Island.

Cox, R.J. 2010. *Environmental Communication and the Public Sphere*. London: Sage.

Crosson, S. 2019. *Gaelic Games on Film*. Cork: Cork University Press.

Cubitt, S. 2005. *EcoMedia*. New York: Radopi.

———. 2017. *Finite Media: Environmental Implications of Digital Technology*. Durham North Carolina: Duke University.

DeLuca, Kevin. 2006. *Image Politics: The New Rhetoric of Environmental Activism*. Mahw Hollywood Utopia: Ecology in Contemporary ah: Erlbaum.

Drezner, D.W. 2011. *Theories of International Politics and Zombies*. Princeton: University of Princeton Press.

Dryzek, J. 1997. *The Politics of the Earth: Environmental Discourses*. Oxford: Oxford University Press.

Garforth, L. 2018. *Green Utopias: Environmental Hope Before and After Nature*. Cambridge: Polity Press.

Hansen, A., and R. Cox, eds. 2015. *Routledge Handbook of Environment and Communications*. Abingdon: Routledge.

Heise, U. 2008. *Sense of Place, Sense of Planet: The Environmental Imagination of the Global*. Oxford: Oxford University Press.

Hickel, Jason, and Giorgos Kallis. 2019. Is Green Growth Possible? *New Political Economy*, April.

Hoggett, P., ed. 2019. *Climate Psychology: On Indifference to Disaster*. Cham: Palgrave.

Hume, M. 2009. *Why We Disagree About Climate Change: Understanding Controversy, Inaction and Opportunity*. Cambridge: Cambridge University Press.

Jackson, T. 2009. *Prosperity Without Growth*. London: Earthscan.

Lovelock, J. 2009. *The Vanishing Point of Gaia*. New York: Basic Books.

Marcu, A., et al. 2015. Analogies, Metaphors, and Wondering about the Future: Lay Sense-Making Around Synthetic Meat. *PUS Public Understanding of Science* 24 (5): 547–562.

Maxwell, R., and T. Miller. 2012. *Greening the Media*. Oxford: Oxford University Press.

Miller, T. 2017. *Greenwashing Culture*. Abingdon: Routledge.

Moore, E. 2017. *Landscape and the Environment in Hollywood Film: The Green Machine*. London: Palgrave.

Morton, T. 2018. *Being Ecological*. Cambridge, MA: MIT Press.

Murphy, P.D. 2017. *The Media Commons: Globalisation and Environmental Discourses*. Chicago: University of Illinois Press.

Negra, D. 2007. Fantasy, Celebrity and "Family Values" in 'High-End and Special Event Tourism in Ireland. In *Irish Postmodernisms and Popular Culture*, ed. W. Balzano, A. Mulhall, and M. Sullivan. Basingstoke/New York: Palgrave Macmillan.

Parham, J. 2016. *Green Media and Popular Culture*. Basingstoke: Palgrave.

Radley, E. 2013. Violent Transpositions: The Disturbing "Appearance" of the Irish Horror Film. In *Viewpoints, Theoretical Perspectives on Irish Visual Texts*, ed. C. Bracken and E. Radley. Cork: Cork University Press.

Rust, S., S. Monani, and S. Cubitt, eds. 2013. *Ecocinema Theory and Practice (AFI Film Readers)*. Abingdon: Routledge.

Shanagher, S., and P. Brereton. 2019. *Pilgrim Hill*: Alienated Farmers and Degraded Ecologies. *Capitalism, Nature Socialism*. https://doi.org/10.1080/10455752.2019.1670710.

Shepherd, R., J. Barnett, H. Cooper, A. Coyle, J. Moran-Ellis, V. Senior, and C. Walton. 2007. Towards an Understanding of British Public Attitudes Concerning Human Cloning. *Social Science & Medicine* 65 (2): 377–392.

Siegrist, M. (2008). Factors influencing public acceptance of innovative food technologies and products. *Trends in Food Science & Technology* 19: 603–608.

Strain, E. 1998. E. M. Foster's Anti-Touristic Tourism and the Sightseeing Gaze of Cinema. In *Postmodernism in the Cinema*, ed. C. Degli-Esposti, 147–166. New York: Berghahn Books.

Tenbült, P., N.K. de Vries, E. Dreezens, and C. Martijn. 2005. Perceived Naturalness and Acceptance of Genetically Modified Food. *Appetite* 45 (1): 47–50.

Weik von Mossner, A. 2017. *Affective Ecologies: Empathy, Emotion and Environmental Narratives*. Columbus: Ohio State University Press.

Willoquet-Marscondi, P., ed. 2010. *Framing the World: Explorations in Ecocriticism and Film*. Charlottesville: University of Virginia Press.

SECTION IV

# Community Engagement, Education and Activism

# Community Engagement and Community Energy

*Clare Watson*

## Introduction

When it comes to climate change and the energy transition, there is a fundamental problem: most people in Ireland are struggling to curb their own greenhouse emissions, and many are resisting renewable energy developments in their area. Yet, in order to meet the climate challenge,

The research for this chapter was carried out as part of the EPA-funded project 2014-CCRP-MS-21 and is published alongside the EPA Research Report (Watson, C., Boyle, E., Mullally, G., Ó Gallachóir, 2019, in press. *Responding to the Energy Transition in Ireland: The Experience and Capacity of Communities Department of Sociology and School of Engineering*. Cork: MaREI Centre, Environmental Research Institute. University College Cork), as part of the EPA Research Programme 2014–2020 (2014-CCRP-MS-21). The programme is financed by the Irish Government. It is administered on behalf of the Department of Communications, Climate Action and the Environment by the EPA, which has the statutory function of co-ordinating and promoting environmental research.

C. Watson (✉)
UCC, Cork, Ireland
e-mail: clare.watson@ucc.ie



D. Robbins et al. (eds.), *Ireland and the Climate Crisis*, Palgrave Studies in Media and Environmental Communication,
https://doi.org/10.1007/978-3-030-47587-1_12

people will need to actively participate in, and support, the substantial lifestyle, infrastructural, and institutional changes that have to be made. This chapter argues that community engagement and, in particular, community energy, is an important part of the solution. But it will not develop effectively without realistic policy support. Initially, it is proposed that a focus on the individual 'rational actor' is misplaced and that, in order to mobilise people, we need to engage them collectively. The concept of energy democracy is introduced, and it is suggested that one way of engaging people in climate action is through community energy groups. The next section outlines the development of community energy policy in Ireland and the role of the Sustainable Energy Authority of Ireland (SEAI), and finally some of the challenges inherent in the community energy sector are explained. Three case studies are included to illustrate the range of activities currently being undertaken by different groups across the country.

## Moving the Focus from the 'Rational Actor' to 'Community Engagement'

In the past, much of the policy focus in relation to climate change mitigation has presumed that individuals make rational decisions based on the information before them—that they weigh up the costs and benefits and then make the choice that appears to be in their own best interest (Jackson 2005). Often the assumption has been that people are 'economically rational' and that an appropriate price signal will stimulate the necessary response. However, this has been shown to be unrealistic and perhaps explains the limited effectiveness of some previous climate action policies (Van Bavel R. et al. 2013). It is now more widely accepted that what many people think they will do, say they will do, and then actually do may differ substantially. In many ways, humans are 'predictably irrational' (Ariely 2008). Therefore, providing individuals with information, on its own, will not result in effective climate action.

For instance, in terms of what people can do to cut their greenhouse emissions, rational economic analysis sees house retrofitting as the obvious 'low hanging fruit'. On the surface, it appears to be a win-win situation—the government tells people what they need to do, offers grants to speed up the process, and gets energy savings in return. The householder makes an initial investment which is repaid over time by reduced energy bills, and comfort levels increase in the home. However, the rate of take-up so far indicates that the situation is more complex than it looks. Despite the fact that making homes more energy efficient saves money, there is an 'energy efficiency gap' (Jaffe and Stavins 1994), with most householders discounting the future benefits. 'In some ways finance is the last barrier people face with regard to energy efficiency' (Hession 2013: 52).

Similarly, the rational choice argument would indicate that, on learning the importance of renewable energy in the fight against climate change, citizens would welcome the development of renewable energy projects. While national opinion surveys generally indicate a high level of support for renewable energy in principle (Upham 2009), it is a mistake to take this for granted and to expect people to welcome developments they claim to support (Wolsink 2000). There is very often a gap between what people say they will accept and what they actually do when faced with a development proposal for their area (Batel et al. 2015). Most people don't think about the desirability of a particular development until a proposal is made to site one in their neighbourhood (Wolsink 1994). Too often, poor governance ensures that residents only learn about a proposed development from a planning notice. If they are concerned, the only active role they can then play is to object. There is a fundamental difference between telling people what development will be taking place within their area and allowing communities to demonstrate what kind of development they find acceptable (Jones and Eiser 2010).

As climate change is such a collective, political, and wide-ranging problem, citizen engagement is crucial, and its absence can leave a vacuum for dissent. Engagement requires that people connect personally, cognitively, emotionally, and behaviourally (Whitmarsh and O'Neill 2012). It is not enough for them to know about the issue, they also need 'to care about it, be motivated and able to take action' (Lorenzoni et al. 2007: 446). Their values, and sense of identity, are then reflected in their actions, that is, *I am the kind of person who does something about climate change* (Climate Outreach 2019).

Such a level of commitment is unlikely to be cultivated with a top-down, individualistic policy approach. After all, people have a fundamental desire and need to belong and to know that they are socially connected to others. Their behaviour is influenced by social norms, social practice, social identity, and fitting in. The thinking now is that, rather than expecting people to make the necessary changes on their own, they need to be mobilised to work on the solution together. Whereas individual actions can appear insignificant and pointless in themselves, when carried out collectively they demonstrate tangible public benefits and a wider social impact (Rogers et al. 2018).

Organised community engagement involves social learning, collective responsibility, and action, with the changes becoming embedded in social norms and practice (Sheppard et al. 2015). Capacity building is crucial. Groups have very different starting points in terms of the knowledge and experience that contribute to effective participation (Head 2007). Communities will have differing skills and access to resources and they may face distinct obstacles (Catney et al. 2014) or lack social cohesion, confidence, and organisational skills (Catney et al. 2013). Therefore, infrastructural and institutional support is essential.

## Energy Democracy and Community Energy

Democratisation involves the inclusion of diverse groups in political affairs and their participation in decision-making and the formation of policy (Dryzek 1996). For democracy to be effective, citizens need to be active and involved both politically and socially (Honohan 2005; Harris 2010). The measure of participation is whether or not citizens are able to gain decision-making power over issues which impact on their lives (Arnstein 1969). The energy democracy agenda seeks to ensure that democracy, citizen participation, and community engagement are at the forefront of the energy transition, that renewable energy systems are planned democratically and are publicly or community owned, and that they deliver tangible benefits to citizens (Burke and Stephens 2018). Energy democracy challenges the techno-economic narrative which sees people as consumers and instead emphasises the involvement of the public as stakeholders (Mullally et al. 2018). It envisages a new kind of *energy citizenship* (Devine-Wright 2004), whereby individuals, co-operatives, and local communities can now invest and benefit from small-scale, distributed renewable energy developments. In so doing, they become 'prosumers', who, while not necessarily being energy self-sufficient, are simultaneously producers and consumers of energy (Szulecki 2018).

In line with energy democracy is the concept of community energy. Community energy involves 'citizen and local ownership and participation in renewable energy generation, distribution and energy efficiency' (Friends of the Earth et al. 2014). It includes 'community projects or initiatives focused on the four strands of reducing energy use, managing energy better, generating energy or purchasing energy' (DECC 2014:

20). The projects or initiatives often arise from the 'grassroots', that is, from the bottom-up by civil society, as opposed to being driven from top-down by governments or other agencies (Klein and Coffey 2016), and they share an emphasis on community ownership, leadership, or control and on community benefits (Oteman et al. 2014: 2). In principle, community energy should create opportunities for all types of communities, beyond the choice few (Catney et al. 2014).

It is generally agreed that the catch-all definition allows for flexibility in relation to approach, participation, and implementation (Hargreaves et al. 2013; Seyfang et al. 2013; Friends of the Earth et al. 2014). It also facilitates experimentation (Walker and Devine-Wright 2008). The lack of any required structure or outcome enables groups to respond to local contexts, conditions and needs, as well as the beliefs and aspirations of their members. Research (Walker and Devine-Wright 2008) demonstrates that projects differ depending on who initiates and runs them, who participates and makes the decisions, and who benefits both socially and, if profits arise, financially. Groups can be non-profit, with charitable status and no business interests, or they could be centred around a community building. Local people may have a financial stake, or shares, or be part of a community co-operative. People who invest and participate in community energy groups are often 'innovators' who are not afraid of risk and experimentation. Many are 'early adopters' who, once they see a clear benefit, enjoy the challenge of trying out new technologies during their growth phase (Bauwens 2016).

The community energy sector began to emerge in the UK in the mid-1990s, with a rise in new groups from 2006 until 2009, followed by a gradual decline (Seyfang et al. 2013). UK government policy began to focus on community-owned generation of renewable energy between 2000 and 2003 (Walker et al. 2007). A 2013 report in the UK stated that over 5000 community groups were involved in energy initiatives during the previous five years (UKGOV 2019). In 2019, Scotland's national community energy charity Community Energy Scotland (CES 2019) claimed to have 400 community group members across the country.

The story is somewhat different here in Ireland. As of 2019, Templederry Community Windfarm, in County Tipperary, is the only community energy project producing electricity for the national grid. The SEAI SEC network includes over 200 communities who are interested in community energy, and some have embarked on energy saving and energy efficiency initiatives in their areas, but few are, as yet, producing renewable energy.

## The Development of Community Energy Policy in Ireland

Policy interest in community energy was first mooted in Ireland in the 1999 Green Paper on Sustainable Energy, which followed the European Commission's 1997 White Paper on Renewable Energy. The Green Paper strongly endorsed the production of renewable energy 'to meet one's own needs' and the development of projects by local co-operatives and other representative organisations (REP 2004: 13).

In Ireland, a country with an abundance of wind, the policy focus was then on wind energy. In early 2000, the state appointed Renewable Energy Strategy Group, produced a *Strategy for Intensifying Wind Energy Deployment*, which noted that part of the challenge of increasing local involvement in wind energy development was that it would involve a significant change in policy direction. Before deciding on options, the report noted it would be useful to 'first decide whether the objective is to reduce the number of objections to large wind farms at the planning stage or to increase local participation in wind energy development' (Fitzgerald 2000: 88).

Essentially, the government needed to decide what it wanted. This sentiment was reflected in March 2000, in a letter to the Irish Times from Séamus Ó Drisceoil, Comhdail LEADER 11 Officer, Oileán Chléire, Cork (O'Drisceoil 2000):

> Comdháil Oileáin na hÉireann [Irish Islands Federation] and others have made repeated submissions to the Green Paper on Energy and elsewhere on the need for continuous access to the grid for small wind-power projects which could be promoted by individuals or communities. Given the right scheme we could have communities embracing wind power on a vast scale rather than uniting to oppose projects. So far absolutely nothing concrete has been achieved in this area.
>
> Here on Oileán Chléire and neighbouring Bere Island we have full planning permission and funding available for small 0.5MW wind projects. We could be in production within six months. This exercise could be repeated throughout the country as communities and farmers see the benefits of wind energy. The technology is tried, tested and absolutely reliable.
>
> So far our access to the grid has been blocked while the Department look to unproven and vastly more expensive technology which is, apparently, to be placed in "someone else's back yard".
>
> Not good enough!

Both the Oileán Chléire and Bere Island wind projects subsequently collapsed.

By 2003, there were only two community-owned wind energy projects in Ireland—three 225 kW turbines on Inis Meáin, County Galway, and a 660 kW turbine installed by the Burtonport fishing co-operative in Count Donegal—which stood in stark contrast to Denmark, where a total of 377 turbines had been erected between 1979 and 1980 and wind power guilds had been set up all over the country, drawing on a rural co-operative tradition similar to that in Ireland (REP 2004). The *To Catch the Wind* report, produced by two County Mayo community wind groups and the statutory Western Development Commission (REP 2004), noted that Danish communities became involved in wind energy at a time when the technology was in its infancy, and the turbines and windfarms were too small to interest large developers, thereby allowing small locally financed community projects to flourish. A significant shift in government policy and a degree of protection was required if Irish communities were to gain a similar share of wind energy development. The report called for a feed-in tariff, free access to the grid, state support and incentives, and a 'one-stop-shop' for community groups needing expert technical, legal, and financial advice. In the absence of progress on this, the advice from the REP to communities was stark: don't invest in wind energy projects 'as the level of risk and uncertainty is currently too high'.

The Government's 2007 White Paper on Energy (DCMNR 2007) made no reference to the development of community involvement in renewable energy projects or the elimination of barriers.

In 2011, the Sustainable Development Council, Comhar, released a report called *Community Renewable Energy in Ireland: Status, Barriers and Potential Options* (Comhar 2011), which reiterated the four main barriers to community renewable energy in Ireland—insufficient policy framework; inadequate support structures; lack of access to finance; and grid and planning delays. This was echoed by the 2012 NESC Report *Ireland and the Climate Change Challenge: Connecting 'How Much' with 'How To'* (NESC 2012), which also highlighted the need for capacity building in the sector.

In 2014, the NESC Report *Wind Energy in Ireland: Building Community Engagement and Social Support NESC* stated that, as part of an inclusive community engagement process to shape and share local value of wind development projects, national policy supports and measures should include 'incentives and measures for promoting community [and] co-operative energy schemes and new financial mechanisms for public investment in renewable energies' (NESC 2014: 5).

Finally, in 2015, the Energy White Paper, *Ireland's Transition to a Low Carbon Energy Future 2015–2030* (DCENR 2015), was published and demonstrated how policymakers were really beginning to take the issue of citizen and community engagement in the energy transition seriously. 'The transition will see the energy system change from one that is almost exclusively Government and utility led, to one where citizens and communities will increasingly be participants in energy efficiency and in renewable energy generation and distribution' (ibid., Chapter 4). The intention to address the challenges and barriers was very clear. Commitments were made to provide funding for community-led projects in the initial stages of development, planning, and construction and to facilitate access to the national grid.

The 2015 Energy White Paper was followed by the report *Assessment of Models to Support Community Ownership of Renewable Energy in Ireland* prepared for SEAI in 2017 (Morris et al. 2017), and in 2018 by the Department's Renewable Electricity Support Scheme (RESS), which promised to 'deliver opportunities for communities, in terms of increased engagement with, participation in, ownership of and benefit from, renewable electricity projects' (Govt of Irl 2018: 2). Far-reaching proposals on how this could be done were included in a public consultation process in January 2020.

## The Role of the Sustainable Energy Authority of Ireland (SEAI)

SEAI is Ireland's national energy authority and is playing a leading role in Ireland's energy transition. Between 2007 and 2013, it led a five-year EU HOLISTIC project focusing on the development of Dundalk as an 'exemplar community' with the aim of stimulating a national move towards sustainable energy practice. Experience gained from this project informed the setting up of SEAI's Better Energy Community (BEC) scheme, which supports innovative energy efficiency projects at community level by offering grants for the retrofitting of local building and development of small-scale renewable energy projects. This is a competitive programme piloted in 2012 and now runs annually. Partnerships are encouraged to apply and might include 'collaborations between public and private sectors, residential and non-residential sectors, commercial and not-for-profit organisations, or financing entities and energy supplier' (SEAI 2019a).

In April 2016, SEAI launched their Sustainable Energy Communities (SEC) Programme, calling on local communities to get involved and join the SEC Network. A Sustainable Energy Community is a 'community in which everyone works together to develop a sustainable energy system for the benefit of their community ... An SEC can include all the different energy users in the community including homes, sports clubs, community centres, churches and businesses'. The SEC Network enables SECs 'to engage and learn from project site visits, seminars, events, and case studies' (SEAI 2019b). SECs are encouraged to enter into a three-year Partnership Agreement with SEAI, involving the production of an Energy Master Plan (EMP) and a Work Plan. Technical and mentoring support is available to each group, along with consultancy costs relating to the EMP and financial supports for developing core competencies to implement the Work Plan. Internal employment is not an eligible cost.

Recognising that people do not act rationally and that behavioural and psychological factors affect the uptake of sustainable energy measures, SEAI established the Behavioural Economics Unit in 2017 to identify and test scalable solutions within Ireland (SEAI 2020).

## The Challenges of Community Energy

Research involving a survey of 190 UK community energy groups (Seyfang et al. 2013) noted that, while they are 'cautiously optimistic' for the development of the community energy sector in the UK, there are inherent tensions in the model. The authors question the ability of groups to scale up and to become more professional and commercial, especially if they continue to operate on a voluntary basis. Seventy-nine percent of the projects surveyed were less than five years old, and the average age of groups was just over four years, which raises certain questions about their long-term viability. The researchers concluded that balancing the needs of members and supporters with the complications of the tasks involved can be difficult without external sources of finance and support and consistent policy backing. The growth potential of voluntary groups is uncertain. The diversity of the sector and its focus also means that government departments need to work together to ensure best performance—the outcomes cannot just be measured in kilowatt hours. Therefore, it is their contention that community energy will not necessarily be a policymaker's shortcut to widespread change.

Similarly, findings from research carried out on two Irish community energy projects (Cogan 2017)—Erris Sustainable Energy, established in 2014 on the north west coast of County Mayo, and the Energy Communities Tipperary Co-operative (ECTC), which began in Drombane, County Tipperary in 2010—suggest that, while community energy initiatives can play a vital role in initiating societal climate action in Ireland, the sector will not flourish without clear political encouragement and realistic support.

Further Irish research (Watson et al. 2019-in press) found that out of 15 'grassroots' community energy projects proposed in Ireland between 1986 and 2010, only three appeared to be operational in 2019—Burtonport Fisherman's Co-op's wind turbine, Ballytobin Camphill's anaerobic digester, and Templederry Community Windfarm (see Table 1). The others did not get off the ground. As part of this study, more in-depth research was carried out with active community energy groups between 2015 and 2018, including Templederry Community Windfarm; Aran Islands Energy Co-op; Claremorris & Western District Energy Co-op; Cloughjordan Ecovillage; Energy Communities Tipperary Co-op; Kerry Sustainable Energy Co-op; and Terenure Sustainable Energy Community. Only Templederry and Cloughjordan Ecovillage were then actively producing energy. The research found that, while energy citizenship is an accepted ambition, energy communities are still struggling. Barriers which existed in 1986, such as the lack of feed-in tariffs and difficulties in gaining planning permission, securing investment finance, and obtaining access to the grid, are still in place and need to be addressed. Community energy does not guarantee community involvement, acceptability, or acceptance. Therefore, mentoring in community development and community engagement is essential, and a range of approaches should be developed in response to the varying capacities of different communities. Intermediary groups should be supported to give advice and assistance, and finally, groups cannot rely solely on volunteers, so the lack of core funding needs to be addressed.

**Table 1** Community energy projects in Ireland, 1986–2010

| *Start date* | *Location* | *Group* | *Aim* | *Actions* | *Funding* | *Challenges* | *End result* |
|---|---|---|---|---|---|---|---|
| 1986 | Cape Clear Island, Co Cork | Cape Clear Co-operative | To develop the first successful variable pitch wind turbines in Ireland and to provide electricity for the island | Two 50 ft. 30 kW turbines were installed on the island | German manufacturers, SMA Regelsystem Gmbh, provided the technology and used project as test-bed | Turbines proved to be uneconomical and required intensive technical servicing; underwater cable bringing electricity from the mainland was installed | Turbines went out of use in 1997 |
| c. 1994, Plans announced in 2009 | Mount Callan, Co Clare | West Clare Renewable Energy Ltd (WCRE)—30 local farm families, with 3000 acres of land; McCarthy Keville O'Sullivan (MCKOS) managed project through EIA and planning process | To install 29 3 MW wind turbines on western slopes of Mount Callan | Progressed through feasibility stage; planning approved by An Bord Pleanála (2011); WCRE partnered with Brookfield Renewable Energy Group | Group was keen that the project be funded by local shareholders, but it appears that this did not occur | Grid connection system, local opposition | Windfarm comprising 11 N90/2500 turbines under construction (2017). Group has committed to funding four local communities, each receiving €100 k initially, then €20 k a year for 5 years, €10,250 annually for next 15 years, and €5 k annually for last 5 years, totalling approx. €378 k for each parish |

(*continued*)

**Table 1** (continued)

| *Start date* | *Location* | *Group* | *Aim* | *Actions* | *Funding* | *Challenges* | *End result* |
|---|---|---|---|---|---|---|---|
| 1995 | Ballytobin, Co Kilkenny | Camphill Community Ballytobin (with 80 residents) set up Bio-Energy & Organic Fertiliser Services (BEOFS) to run the project; four people employed to operate the plant | To build an anaerobic Digestion/ Biogas Plant for the Ballytobin Camphill community; to create work for residents and demonstrate centralised anaerobic digestion for first time in Ireland | Construction began (1996); project began fuelling a small district heating system (1999), using slurry from local farms and food waste from waste management companies | Investment by Camphill Ballytobin; Camphill Community real estate used for bridging loans; Rural Dev. Prog.; EU Leader Prog. II; EU 'Horizon'; EU 'Altener'; gate fees funded two employees, two CE Scheme employees | Accessing capital funding; inability to obtain Power Purchase Agreement to connect plant to grid so, in warmer months, excess biogas had to be flared off | Ballytobin was one of nine Camphill sites to benefit from SEAI BEC (2015) upgrades, which included a biogas CHP plant to generate electricity |
| 1997 | Cape Clear Island, Co Cork | Cape Clear Community Council | | Feasibility study for RE trail; interim report on energy conservation, recycling, waste mgmt and wind developments; enviro. Reports on proposed upgrading of wind energy system; potential for other RE projects investigated; PP granted 0.5 MW wind turbine; two energy managers trained | EU Partnership project under Regional and Urban Energy Planning Programme; Cork Co Co; Udaras Na Gaeltacht; LEADER | Accessing the grid | Project ended; wind turbine was not erected |

| | | | | | | | |
|---|---|---|---|---|---|---|---|
| Late 1990s | Inis Meáin, Co Galway | Inis Meáin Island Co-op | To create electricity to power desalination plant | Three Vestas V27 225 kW wind turbines installed to power a new desalination plant (2002) | EU—Fifth Framework; Údarás na Gaeltachta; Galway Co Co | Enviro. groups objected to original planning application; local co-op became mired in controversy and subsequently disbanded | 2011, the desalination plant closed down; turbines fell into disuse, despite efforts to bring them back into operation were dismantled for safety reasons |
| c. 1999 | Bere Island, Co Cork | Wind energy co-op, with 200 island residents and part-time residents as €1 shareholders; one person worked on project for 18 months | To install 600 kW Vestas wind turbine, linked to mainland grid; to use profits for island development projects | Obtained Power Purchase Agreement (AER 5) and planning permission | €100,000 raised from island sources | Failed to secure EU INTERREG and other funding; group unable to secure the €200,000 necessary for project viability; process very complex | Group lost momentum; project shelved (2003); turbine planning expired (2004) |
| 1999 | Freshford, Co Kilkenny | The 'Freshford Alive' formed by Freshford 2020 development group; reps of BNS Leader, Kilkenny Co Co and Tipperary Inst. project steering committee; full-time consultant co-ordinator hired | Address village sewerage system sustainably, using local waste for CHP plant producing electricity for grid and gas for local heating, and to provide secondary sewage treatment using water hyacinths | Feasibility study and development plan produced | EPA, SEI, and LEADER (€20,000 for feasibility study); EU INTERREG (€41,799 for development phase) | | Project appears not to have progressed; Freshford 2020 Rural Dev. Ltd dissolved sometime after Jan 2006 |

*(continued)*

**Table 1** (continued)

| *Start date* | *Location* | *Group* | *Aim* | *Actions* | *Funding* | *Challenges* | *End result* |
|---|---|---|---|---|---|---|---|
| PP granted in 2000 | Ballycogley, Co Wexford | Wexford Wind Energy Co-op, in partnership with developer | To install four 3.5 MW turbines on a 150-acre site at Ballycogley, with two turbines financed by developer, and shares for other two to be offered to local community with preference for those closest to site | Progressed through feasibility stage; planning permission granted in 2000 | EU THERMIE grant; hoped to raise remaining funds through corporate tax relief scheme | High grid connection costs | Project did not proceed; Ballycogley Wind Energy Plc dissolved in 2007 |

| | | | | | | | |
|---|---|---|---|---|---|---|---|
| 2002 | Killala, Co Mayo | Killala Community Wind Farm Ltd (8 farmers, 3 directors, and 17 shareholders), in partnership with Killala Community Council (KCC), with assistance from Western Dev. Commission (WDC); WDC assigned rural development worker to work on project | To develop a 23 MW community windfarm and encourage local people to invest through a number of 'investment vehicles' | Project team (two KCWF directors, KCC dev. manager, two KCC members, WDC rural dev. worker) (2006); WDC provided initial project co-ordination, facilitation, technical and management expertise, assisted with provision of information to public; planning application submitted (2007); 45 people objected; An Bord Pleanála refused permission; PP granted (six turbines, 2010) | Farmers provided initial seed funding; SEI (feasibility phase and €39,000 to document how local communities can become involved in wind energy); WDC | Lack of explicit policy supports; complexity of RE projects; negative media coverage of wind; length of process strained community resources and entrenched 'anti' positions; difficult to demonstrate benefits to wider community; difficult to identify appropriate inclusive and representative community; difficult to communicate between parties | Killala Renewable Production Limited ('KRPL') (parent company of KCWL) and Gaelectric Developments Ltd joined forces (2015) and applied for modifications to 2010 permission (2017); applicant intends to give €1000 per MW to a community fund each year |
| 2003 (operational) | Burtonport, Co Donegal | Burtonport Fishermens Co-Op | To provide electricity for fish ice plant | One Vestas V47 660 kW wind turbine installed | | | This turbine remains in operation |

*(continued)*

**Table 1** (continued)

| *Start date* | *Location* | *Group* | *Aim* | *Actions* | *Funding* | *Challenges* | *End result* |
|---|---|---|---|---|---|---|---|
| 2006 | Co Waterford | Waterford Renewable Energy Co-operative Society Ltd. (established by Waterford Co Co & Waterford Energy Bureau) | To be a pilot rural self-supply co-operative and develop a number of RE initiatives (bioenergy and wind) for the benefit of its members | Co-op secured 52 members; was facilitating the development of bioenergy projects and three community wind farms (2012) | Energy Self-Supply in Rural Communities (ENSRC) supported by Intelligent Energy Europe (IEE) | | No more online info on this group; their website has been disabled |
| 2008 | Kinsale, Co Cork | Transition Town Kinsale | Sell electricity to national grid, generate heat for use locally, and use bio-waste as agricultural fertiliser | To develop a community-run anaerobic digester, converting local farm/food waste into locally used energy | West Cork Dev. Partnership (€10,000 to determine project viability); Rethink, Recycle, Remake (Rx3) programme | Finding site; lack of interest by locals in home heating option; changed focus to providing gas for local vehicles, but local farmers not interested in any capacity | The project is currently dormant |
| 2010 | Ballylaneen, Stradbally, & Bunmahon, Co Waterford | BSB Community Energy Ltd, established by two local landowners, with local committee, 50 local shareholders | To erect 11 wind turbines producing up to 33 MW of electricity and to set up a community-owned company | Investors were acquired and plans progressed over next four years, but there was no public consultation | | Local opposition group 'Mahon Valley Against Turbines'; protest meeting held in Nov 2016 with over 600 attendees | Community energy consultant appointed; public meeting held, too late as strong opposition mobilised (July 2017) BSB withdrew wind farm proposal (Aug 2017); deep divisions locally |

| | | | | | | | |
|---|---|---|---|---|---|---|---|
| 2010 | Ballynagran, Co Wicklow | Ballynagran Community Energy Plus Project run by Zero Carbon Ltd; project manager worked on project | To become the world's first Zero Carbon Community within 15 years, by reducing energy use, creating an energy-independent region, producing RE locally, creating sustainable local employment and enhancing quality of life | Carried out local energy audits; substantial number of local houses retrofitted | Interreg IVD North West Europe; Zecos Project (Zero CO2 Emission Certification System); Wicklow Co Co; Greenstar Ballynagran Landfill Community Fund; Ballynagran Environmental Community Projects & Works Grant Scheme; SEAI; company donations; savings by bulk buying | Unsuccessfully applied to become one of SEAI's SECs (2011); local objections to wind turbine proposal; high degree of complexity; lack of organisational experience and specialist skills; high capital costs of some schemes; financial risks involved; planning permission and planning delays; lack of interest and mistrust | PP granted by Wicklow Co Co for 500 KW wind turbine (2015); proposal invoked local objections; An Bord Pleanála refused permission (2016) due to absence of 'an overall strategy for the development of wind energy in this area… it is considered that the provision of a single wind turbine would represent a haphazard and uncoordinated approach'; this damaged group morale; current status unclear |

## Case Studies

**Templederry Community Windfarm**

The idea for a community-owned windfarm in the small rural townland of Templederry, County Tipperary, emerged after a local development plan highlighted renewable energy options. The project was coordinated by a small local committee, which established Templederry Energy Resources Ltd in 2003. Twenty-eight shareholders were recruited, with two shares going into a community co-operative fund. Templederry Windfarm Ltd was formed in 2010 to deal with financing and power purchase issues. Planning permission was granted by Tipperary County Council. However, this expired before construction, due to a moratorium on grid connections and a two-year delivery delay on turbines. A second planning application was supported by the council, but local objectors appealed the decision to An Bord Pleanála, which, after a delay of over two years, finally granted permission. In 2012, two 2.3 MW turbines were erected and the community windfarm was officially opened by the Ministers for Energy and Environment in 2013. The two turbines currently produce approximately 15 GWh per annum of green electricity, powering the equivalent of 3000 homes. The operation employs one full-time worker. A proposal for a second phase of the community windfarm was objected to locally, and planning was refused by Tipperary County Council and An Bord Pleanála. Grid applications have since been lodged by the group for four solar farms, one in partnership with Claremorris & Western District Energy Co-op.

In 2015, Community Renewable Energy Supply Company (CRES) was established by the group as Ireland's first community-owned licensed supply company trading on the Integrated Single Electricity Market (ISEM). CRES, trading as Community Power, purchases and sells electricity on behalf of its customers and is working to support and promote local energy markets within communities where citizens and communities actively participate in renewable energy generation and distribution projects. In early 2016, Community Power began supplying renewable electricity to local residents in Templederry and the surrounding area. It has since developed its customer base and is now supplying electricity to

(*continued*)

**(continued)**

domestic customers, community-owned and managed facilities, and commercial and local businesses throughout Ireland. If granted a 'Gas Supply Licence' and 'Gas Shipping Licence', the company can, using its current structure, encourage communities to generate and supply biogas for consumption locally and throughout the national network. CRES has one part-time and two full-time employees.

**Aran Islands Energy Co-op**

Established in 2012 through Comharchumann Forbartha Árann (Aran Development Cooperative), Comharchumann Fuinnimh Oileáin Árann (Aran Islands Energy Co-operative) aims to secure energy independence for the three Aran Islands off the coast of County Galway, by 2022. Life membership is €100 and is open to all island residents. Out of a population of about 1200, approximately 100 have so far joined up. By 2019, 250 homes and community buildings had been retrofitted and over 50 heat pumps, 35 PV systems, 12 electric cars, a Tesla battery, a DACS battery, LED lighting, and energy monitoring had been introduced. There has been a 24% reduction in imported heating fuel.

The islands have benefitted from SEAI BEC schemes as well as through their participation in EU-funded projects. The group is one of SEAI's Sustainable Energy Communities (SECs) and is keen to progress its community wind energy proposal. Local concerns have meant that the range of potential sites is limited but the project is still going ahead. The co-op is also looking at plans for solar and marine energy.

The Energy Co-op is working alongside NUIG on a number of EU-funded projects: SEAFUEL—investigating the use of hydrogen as a transport fuel in island communities; RESPOND—researching the benefits of monitoring equipment in homes to improve energy efficiency; and GEOFIT—researching more efficient ways of using geothermal heating in homes. The Energy Co-op is also a participant in Community Power. It employs two permanent staff and two temporary part-time workers.

**Dingle Peninsula 2030**

Dingle Peninsula 2030 is a multi-partner initiative based on the Dingle Peninsula, in County Kerry, working with the local community, schools, and business and farming sectors to explore, support, and enable the broader societal changes required for the low-carbon transition. Established in early 2018, it involves the Dingle Creativity and Innovation Hub, ESB Networks, the MaREI Centre, and North East West Kerry Development (NEWKD). The Dingle Hub has set up the Dingle Sustainable Energy Community (SEC), completed an Energy Master Plan for the area, and is working on the following: a feasibility study for an anaerobic digester; a sustainable transport initiative with Local Link Kerry; and, in partnership with Kerry Education and Training Board, the training of ten local energy mentors to support people in retrofitting their homes and using energy efficiently. The Hub is also hoping to set up a local energy Co-op to enable the community to invest in future renewable energy generation.

ESB Networks (ESBN) is running a €5 million three-year project on the peninsula trialling a range of technologies required for the development of the low-carbon electricity network of the future, including solar photovoltaics, battery management systems, air source heat pumps, and peer-to-peer trading devices. As part of this project, the homes/businesses of five local Ambassadors have been retrofitted and ESBN is trialling battery technology, solar panels, and electric cars with other local residents.

NEWKD in collaboration with MaREI has run a number of community engagements across the seven parishes covering the Dingle Peninsula in order to develop an evidenced-based plan for the sustainable development of the Dingle Peninsula. The Dingle Peninsula 2030 team has an active presence at the many festivals across the region, and they engage with the local media. Work is ongoing with the local secondary schools (e.g. Climate Hack) as part of a major programme on 'Activating the Energy Citizen'.

A team of engaged researchers from MaREI is offering guidance and research support. The evaluation of the project, and lessons learned, will be documented in a case study and shared with relevant stakeholders in order to inform future climate action.

## Conclusion

This chapter recognises that in order to effectively address climate change and to implement the energy transition, people need to be engaged collectively and involved in climate action at all levels. We can no longer presume that individuals act rationally on being presented with the facts. The concept of energy democracy envisages a new kind of *energy citizenship* whereby citizens are simultaneously producers and consumers of energy, and this may mean that they become active in community energy groups. However, community energy is not easy and it will not thrive without realistic policy support and resources.

In Ireland, the development of policy around community energy has been extremely slow. While practitioners were palpably excited by the content of the 2015 Energy White Paper, there has been little concrete progress, particularly around the elimination of barriers to creating community energy and the provision of multi-annual core funding. However, policy shifts in 2018 and 2019, including the launch of the All of Government Climate Action Plan 2019, and the ongoing consultation process around the RESS are very encouraging signs that change is on the way. Nevertheless, while infrastructural supports are emerging, it is really important that they respond more effectively to community needs and that they allow for new possibilities and learning around how to upscale projects. This will require a new flexible form of governance which allows for exploration, experimentation, and cross-fertilisation.

## Bibliography

Ariely, D. 2008. *Predictably Irrational*. New York: Harper Collins.

Arnstein, S.R. 1969. A Ladder of Citizen Participation. *Journal of the American Planning Association* 35 (4): 216–224.

Batel, S., P. Devine-Wright, L. Wold, H. Egeland, G. Jacobsen, and O. Aas. 2015. The Role of (de-)Essentialisation Within Siting Conflicts: An Interdisciplinary Approach. *Journal of Environmental Psychology* 44: 149–159.

Bauwens, T. 2016. Explaining the Diversity of Motivations Behind Community Renewable Energy. *Energy Policy* 93: 278–290.

Burke, M.J., and J.C. Stephens. 2018. Political Power and Renewable Energy Futures: A Critical Review. *Energy Research & Social Science* 35: 78–93.

Catney, P., A. Dobson, S.M. Hall, S. Hards, S. Mac Gregor, Z. Robinson, M. Ormerod, and S. Ross. 2013. Community Knowledge Networks: An Action-Orientated Approach to Energy Research. *Local Environment* 18 (4): 506–520.

Catney, P., S. MacGregor, A. Dobson, S.M. Hall, S. Royston, Z. Robinson, M. Ormerod, and S. Ross. 2014. Big Society, Little Justice? Community Renewable Energy and the Politics of Localism. *Local Environment* 19 (7): 715–730.

CES. 2019. *Interactive Map*. Available at: https://www.communityenergyscotland.org.uk/index.asp

Climate Outreach. 2019. *Mainstreaming Low Carbon Lifestyles*. Available at: https://climateoutreach.org/resources/mainstreaming-low-carbon-lifestyles/

Cogan, S. 2017. Community-led Energy Initiatives in Ireland: Accelerating the Energy Transition? Reflections on the Impact and Outcomes of Two Case Studies. *Irish Journal of Social, Economic and Environmental Sustainability* 1 (1): 4–21.

Comhar. 2011. *Community Renewable Energy in Ireland: Status, Barriers and Potential Options*. Comhar Sustainable Development Council. Available at: http://files.nesc.ie/comhar_archive/Comhar%20Papers/Comhar_Paper_11_2011.pdf

DCENR. 2015. Ireland's Transition to a Low Carbon Energy Future 2015–2030, Department of Communications, Energy and Natural Resources, Dublin. Available at: https://www.dccae.gov.ie/documents/Energy%20White%20Paper%20-%20Dec%202015.pdf.

DCMNR, Department of Communications, M.a.N.R. 2007. *Energy White Paper – Delivering a Sustainable Energy Future for Ireland*. Dublin: DCMNR.

DECC. 2014. Community Energy Strategy: Full Report. Available at: https://assets.publishing.service.gov.uk/government/uploads/system/uploads/attachment_data/file/275163/20140126Community_Energy_Strategy.pdf.

Devine-Wright, P. 2004. Towards zero-carbon: Citizenship, Responsibility and the Public Acceptability of Sustainable Energy Technologies. *Proceedings of Conference C81 of the Solar Energy Society, UK Section of the International Solar Energy Society*, 51–62.

Dryzek, J.S. 1996. Political Inclusion and the Dynamics of Democratization. *American Political Science Review* 90 (3): 475–487.

Fitzgerald, J. 2000. *Strategy for Intensifying Wind Energy Deployment*. Dublin: Government of Ireland.

Friends of the Earth, et al. 2014. Community Energy Policy Position Paper. Available at: https://www.foe.ie/download/pdf/community_energy_policy_position_paper.pdf.

Government of Ireland. 2018. *Renewable Electricity Support Scheme (RESS)*. Dublin: Government of Ireland.

Hargreaves, T., S. Hielscher, G. Seyfang, and A. Smith. 2013. Grassroots Innovations in Community Energy: The Role of Intermediaries in Niche Development. *Global Environmental Change* 23 (5): 868–880.

Harris, C. 2010. Active Democratic Citizenship and Service-Learning in the Postgraduate Classroom. *Journal of Political Science Education* 6 (3): 227–243.

Head, B.W. 2007. Community Engagement: Participation on Whose Terms? *Australian Journal of Political Science* 42 (3): 441–454.

Hession, B. 2013. *Better Energy Financing: Barriers to the Uptake of a National Retrofit Scheme*. Working Document, Better Energy Financing (BEF). Dublin: DCENR.

Honohan, I. 2005. Active Citizenship in Contemporary Democracy. In *The Report of the Democracy Commission: Engaging Citizens, the Case for Democratic Renewal in Ireland*, ed. C. Harris, 169–180. Dublin: TASC at New Island.

Jackson, T. 2005. *Motivating Sustainable Consumption: A Review of Evidence on Consumer Behaviour and Behavioural Change: A Report to the Sustainable Development Research Network*. Guildford: Centre for Environmental Strategy, University of Surrey.

Jaffe, A.B., and R.N. Stavins. 1994. The Energy-Efficiency Gap What Does it Mean? *Energy Policy* 22 (10): 804–810.

Jones, C.R., and J.R. Eiser. 2010. Understanding 'Local' Opposition to Wind Development in the UK: How Big Is a Backyard? *Energy Policy* 38 (6): 3106–3117.

Klein, S.J., and S. Coffey. 2016. Building a Sustainable Energy Future, One Community at a Time. *Renewable and Sustainable Energy Reviews* 60: 867–880.

Lorenzoni, I., S. Nicholson-Cole, and L. Whitmarsh. 2007. Barriers Perceived to Engaging with Climate Change Among the UK Public and their Policy Implications. *Global Environmental Change* 17 (3): 445–459.

Morris, S., J. Harvey, T. Bruton, and G. Vaughan-Morris. 2017. *Assessment of Models to Support Community Ownership of Renewable Energy in Ireland, Report for SEAI*. Didcot: Ricardo Energy & Environment, UK.

Mullally, G., N. Dunphy, and P. O'Connor. 2018. Participative Environmental Policy Integration in the Irish Energy Sector. *Environmental Science & Policy* 83: 71–78.

NESC. 2012. *Ireland and the Climate Change Challenge: Connecting 'How Much' with 'How To'; Final Report of the NESC Secretariat to the Department of Environment, Community and Local Government*. Dublin: National Economic & Social Council. Available at: http://www.nesc.ie/assets/files/Ireland%20and%20the%20Climate%20Change%20Challenge_Connecting%20How%20Much%20with%20How%20To_Main_Report.pdf.

———. 2014. *Wind Energy in Ireland: Building Community Engagement and Social Support*. Dublin: National Economic & Social Council. Available at: http://files.nesc.ie/nesc_reports/en/139_Wind_Energy_Main_Report.pdf.

O'Drisceoil, S. 2000. Energy From Wind Turbines. *The Irish Times*. Available at: https://www.irishtimes.com/opinion/letters/energy-from-wind-turbines-1.252984

Oteman, M., M. Wiering, and J.-K. Helderman. 2014. The Institutional Space of Community Initiatives for Renewable Energy: A Comparative Case Study of the Netherlands, Germany and Denmark. *Energy, Sustainability and Society* 4 (1): 11.

REP. 2004. *Renewable Energy Partnership: To Catch the Wind: The Potential for Community Ownership of Wind Farms in Ireland*. Available at: https://www.wdc.ie/wp-content/uploads/reports_To-Catch-the-Wind.pdf

Rogers, T., N.J. Goldstein, and C.R. Fox. 2018. Social Mobilization. *Annual Review of Psychology* 69: 357–381.

SEAI. 2019a. *Community Grants*. Available at: https://www.seai.ie/grants/community-grants/.

———. 2019b. *Sustainable Energy Communities (SEC) Partnership Agreement Application Guidelines*. Available at: https://www.seai.ie/publications/SEC-Partnership-Application-Guidelines.pdf

———. 2020. *Behavioural Insights*. Available at: https://www.seai.ie/data-and-insights/behavioural-insights/

Seyfang, G., J.J. Park, and A. Smith. 2013. A Thousand Flowers Blooming? An Examination of Community Energy in the UK. *Energy Policy* 61: 977–989.

Sheppard, S., D.M. Iype, S. Cote, and J. Salter. 2015. *Special Report–A Synthesis of PICS-Funded Social Mobilization Research*. Available at: https://pics.uvic.ca/sites/default/files/Soc.%20Mob.%20Report%20_0.pdf

Szulecki, K. 2018. Conceptualizing Energy Democracy. *Environmental Politics* 27 (1): 21–41.

UKGOV. 2019. *Community Energy Guide*. Available at: https://www.gov.uk/guidance/community-energy

Upham, P. 2009. Applying Environmental-Behaviour Concepts to Renewable Energy Siting Controversy: Reflections on a Longitudinal Bioenergy Case Study. *Energy Policy* 37 (11): 4273–4283.

Van Bavel, R., B. Herrmann, G. Esposito, and A. Proestakis. 2013. *Applying Behavioural Sciences to EU Policymaking*. JRC Working Papers JRC83284. Seville: Joint Research Centre.

Walker, G., and P. Devine-Wright. 2008. Community Renewable Energy: What Should it Mean? *Energy Policy* 36 (2): 497–500.

Walker, G., S. Hunter, P. Devine-Wright, B. Evans, and H. Fay. 2007. Harnessing Community Energies: Explaining and Evaluating Community-Based Localism in Renewable Energy Policy in the UK. *Global Environmental Politics* 7 (2): 64–82.

Watson, C., E. Boyle, G. Mullally, and B. O'Gallachoir. 2019. *Responding to the Energy Transition in Ireland: The Experience and Capacity of Communities Department of Sociology and School of Engineering*. Cork: MaREI Centre, Environmental Research Institute. University College Cork, in press.

Whitmarsh, L., and S. O'Neill. 2012. Opportunities for and Barriers to Engaging Individuals with Climate Change. In *Engaging the Public with Climate Change: Behaviour Change and Communication*, ed. L. Whitmarsh, I. Lorenzoni, and S. O'Neill, 10–14. Abingdon: Routledge.

Wolsink, M. 1994. Entanglement of Interests and Motives: Assumptions behind the NIMBY-Theory on Facility Siting. *Urban Studies* 31 (6): 851–866.

———. 2000. Wind Power and the NIMBY-Myth: Institutional Capacity and the Limited Significance of Public Support. *Renewable Energy* 21 (1): 49–64.

# Climate Change Education in Ireland: Emerging Practice in a Context of Resistance

*Fionnuala Waldron, Benjamin Mallon, Maria Barry, and Gabriela Martinez Sainz*

## Introduction

On Friday, 20 September 2019, thousands of children and young people from primary and post-primary schools across Ireland, as part of a wider global movement across 150 countries, staged collective protests seeking government action in the face of accelerating climate change (*Irish Times* 2019). These protests highlighted young people's determination for climate action but also thrust into the spotlight those working within the Irish education system, with many teachers and school principals involved in the organisation, mediation and, in some cases, curtailment of student involvement in these climate protests. The question of whether, and, if so,

F. Waldron (✉) • B. Mallon • M. Barry
Dublin City University, Dublin, Ireland
e-mail: fionnuala.waldron@dcu.ie; benjamin.mallon@dcu.ie; maria.barry@dcu.ie

G. Martinez Sainz
University College Dublin, Dublin, Ireland
e-mail: gabriela.martinezsainz@ucd.ie



D. Robbins et al. (eds.), *Ireland and the Climate Crisis*, Palgrave Studies in Media and Environmental Communication,
https://doi.org/10.1007/978-3-030-47587-1_13

how, the education sector should facilitate young people's involvement in climate action entered the public discussion.

Climate change has been defined as a 'wicked problem' necessitating complex and multi-faceted solutions (Incropera 2015). Education is viewed as an integral part of this response, with the capacity to contribute towards mitigation of the causes of climate change as well as adaptation to its consequences (Selby and Kagawa 2012; Mochizuki and Bryan 2015). However, this educational space is also acknowledged as both complex and contested (Lehtonen et al. 2019), presenting educators with the challenge of addressing difficult knowledge in a politicised and, at times, divisive context. This chapter charts the shifting conceptualisations of Climate Change Education (CCE) within formal education in Ireland and elsewhere, from an individual to a necessarily collective concern, with a focus on children, teachers and policymakers. After exploration of current and emerging practices in CCE, the chapter considers the Irish policy context for CCE, including the underpinning ideologies and inherent contradictions. With reference to two case studies of high-profile events which shape the Irish context for CCE and young people's climate action, the chapter examines the demands faced by teachers addressing the climate crisis and considers the challenges faced by children grappling with the complexity and emotion of climate change.

## Past, Current and Emerging CCE Practice

While early approaches to CCE focused predominantly on developing scientific knowledge of the processes underpinning climate change and its environmental consequences, there was growing awareness that scientific and environmental knowledge in itself did not necessarily lead to pro-environmental attitudes and behaviour in children (Brownlee et al. 2013) or adults (Dijkstra and Goedhart 2012). More comprehensive approaches followed, particularly from the fields of Environmental Education (EE) and Education for Sustainable Development (ESD), as education was recognised as key to increasing climate change literacy and fostering behavioural changes (Mochizuki and Bryan 2015; Waldron et al. 2016). ESD is seen as a holistic and learner-centred approach to education which is participatory, action-oriented and potentially transformative (UNESCO 2016). Adopted by UNESCO as an educational framework at the International Conference on Environmental Education in Thessaloniki in 1997 and promoted by the United Nations Decade of ESD (2005–2014),

ESD gave an increased focus to climate change as a result of the Bonn Declaration arising from the 2009 UNESCO World Conference on ESD, which called for CCE to be placed higher on the international agenda (UNESCO 2009). Long-standing critiques of ESD have argued that the term itself encompasses contradictions and ambiguities, leading to 'definitional haziness' at policy level (Selby and Kagawa 2010: 38) and a failure to question dominant narratives of growth and consumerism (Selby and Kagawa 2010; Hicks 2014). Thus, ESD can support a wide range of practices, from those which promote individualised pro-environmental action to those which exhibit a more transformative intent.

From an Irish perspective, education addressing climate change focused historically on promoting pro-environmental citizenship, encouraging responsible environmental behaviour at individual and school level, rather than on system critique or collective political action. While the participation of young people in school-based decision-making and planning has been a strong characteristic of Irish environmental education practice, particularly through the Green Schools movement, actions tend to be conceptualised as local, personal and safe (Waldron et al. 2014, 2016). Recent efforts to present more critical educational forms, however, suggest that CCE, in the context of ESD, has a vital role to play in prompting the necessary 'transformative shifts in how we think and act' (Mochizuki and Bryan 2015: 4).

CCE has been defined as preparing young people 'for a rapidly changing, uncertain, risky and possibly dangerous future' (Stevenson et al. 2017). This recognition has led to a move away from a focus on 'rear view mirror' approaches (i.e. focusing on questions, problems and responses, as prevalent in the 1980s and 1990s) (Selby and Kagawa 2010) to encompass other dimensions of knowledge and ways of knowing. While current approaches to CCE include the two broad dimensions of mitigation and adaptation at both the local and global level, these dimensions are viewed as complex and multi-faceted. CCE for mitigation focuses on the development of knowledge, skills and dispositions to identify the causes and consequences of climate change and to promote individual and collective action. CCE for adaptation, on the other hand, focuses on the knowledge and skills to manage current risks, reduce general vulnerability and prepare and respond to climate-related hazards, fostering an adaptive capacity to cope with the imminent impact of climate change (Selby and Kagawa 2012; Mochizuki and Bryan 2015; Stevenson et al. 2017). This includes recognition of the differential impact of climate change within and across countries, the inverse relationship between culpability and vulnerability (in that historical responsibility for generating emissions lies in industrialised countries while those countries least responsible are most vulnerable) and the increased risk for those communities and individuals within countries who are marginalised and impoverished. CCE, then, is multi-dimensional and, to a large extent, open-ended. It encompasses approaches which address the emotional and aesthetic dimensions of climate change, enabling young people to deal with uncertainty and ambiguity, to understand that all things are connected, to reconnect with place and with the wonder, beauty and value of the natural world. Selby (2017b: 17), for example, argues for vernacular learning which enables learners to connect with the natural world through learning experiences that are 'locally grounded'. There is also an increasing recognition of the importance of enabling children to question the ideological and economic drivers of climate change and to understand issues relating to climate justice, including issues of migration and conflict, in ways that promote system-level thinking and collective action (Selby and Kagawa 2015). Several studies advocate for approaches to CCE which are participatory, interdisciplinary and premised on creative modes of engagement and which acknowledge the ethical, social and political dimensions of climate change. Such approaches promote critical thinking, engage with the emotional complexities of climate change and move away from top-down processes to listen to the voices of young people (Mochizuki and Bryan 2015; Rousell and Cutter-Mackenzie-Knowles 2019). The capacity for transformative models of CCE to take root in a state-governed education system, however, is dependent on a range of factors, including the preparedness of schools and teachers to embrace such approaches. It is important, then, to consider the policy environment which shapes teacher practice in Ireland.

## Systemic Forces and Counter Subversions Within Irish Educational Policy

Since the 1960s, Irish educational policy has been framed predominantly by the needs of the economy, neglecting other more social and civic perspectives. This has led to an over-reliance on technical solutions and a focus on issues such as performativity (e.g. testing and target-setting), school infrastructure and top-down reform (Ball 2003; Gaynor 2013; Lynch et al. 2012). Policymaking in the context of Irish education has also been regarded as a-theoretical (O'Sullivan 1989; Lynch et al. 2012), an approach which O'Sullivan (1989) viewed as problematic as it replaces principles rooted in social theory, with morally loaded slogans. Consequently, contradictory slogans (e.g. 'sustainability' and 'economic growth') can coexist within the same educational policy, though these contestations can be masked (O'Sullivan 1989). Furthermore, it is maintained that education policy in Ireland has been characterised by a drive towards consensus and a reluctance to disturb the status quo (Lynch et al. 2012). This is exemplified in the consultative approach taken to curricular reform, where consultation with a range of representative bodies, including teacher unions and employers' groups, makes compromise inevitable (Lynch et al. 2012).

Despite these tensions, aspects of curricula, at both primary and second level, can be viewed as open to more transformative conceptualisations of CCE. Since 1971, the primary curriculum, for example, has been characterised by a holistic, child-centred approach to teaching and learning, and a view of children as active agents in the construction of knowledge (Waldron 2004), while recent curricular reform at post-primary level has afforded increased recognition to active learning methodologies and to the concepts of sustainability, well-being and student voice (NCCA 2019). For example, statements of learning within the Framework for Junior Cycle (NCCA 2015, p. 12) include stipulations that students will 'value what it means to be an active citizen with rights and responsibilities in local and wider contexts' and that students will 'have the awareness, knowledge, skills, values and motivation to live sustainably'. In addition, Aistear, the curricular framework for early childhood education in Ireland, includes children's citizenship as a core principle (NCCA 2009). Subsumed under the idea of 'sustainability', opportunities to integrate CCE were identified by a recent ESD audit conducted by the National Council for Curriculum and Assessment (NCCA 2018) across Irish curricula at preschool, primary level (subjects such as History, Geography, Science and Social, Health and Personal Education) and post-primary level (Business Studies, Home Economics and Visual Art) (NCCA 2018). However, while these characteristics suggest scope to integrate CCE into current practices, as yet, no mandatory CCE obligations exist. Indeed, the NCCA (2018) recognises that 'the scope for addressing content related to specific Sustainable Development Goals is largely dependent upon the professional capacity, interest and disposition of the teacher' (2018: 89).

Ireland's policy on CCE is part of a wider international framework which requires states to respond to climate change through education and offer policy supports to shape that response. Obligations arise from the Sustainable Development Goals and the UN Framework Convention on Climate Change (UNFCCC). CCE is also supported by international policy initiatives such as the OECD's (2018) Global Competence Framework and UNESCO's publications on Global Citizenship Education (2015) and ESD (2017). Such wider initiatives have been heavily critiqued from an educational perspective as lacking the transformative intent required to challenge the economic growth models which continue to drive climate change (Selby 2017b). The Irish National Strategy on Education for Sustainable Development (DES 2014), launched as part of Our Sustainable Future (DECLG 2012), is open to similar critique in terms of its framing of ESD and short fallings at the level of implementation. While there are some examples of more radical and critical programmes across the formal education sector, responses to them demonstrate the controversial nature of CCE and its potentially subversive status. For example, reactions to an educational resource produced by An Taisce (see Case Study 1) demonstrate the potential for CCE to unearth vested interests that the consensual approach in educational policymaking seeks to avoid. This case also highlights the resistance policymakers can encounter when seeking to enable teachers and students to address complex and critical questions related to sustainability, such as the political impetus towards economic growth. This results in significant contradictions within educational policy relating to climate change and, ultimately, a failure to provide a more critical, state-led approach to CCE.

The role of education within wider government climate action policy is increasingly recognised, though recommendations are underdeveloped (DCCAE 2019; Joint Oireachtas Committee on Climate Action 2019). Seventeen references are made to education in the Climate Action Plan (DCCAE 2019), with the majority relating to the Energy in Education Programme or the role of education in supporting workers to develop new, future-oriented skills as part of a 'just transition'. A smaller number of references are made to teaching and learning opportunities, such as embedding climate science in the curriculum and supporting student activism. There are no references to more radical forms of CCE that necessitate an interrogation of current economic models or justice-oriented mitigation strategies. Moreover, within the report from the Joint Oireachtas Committee on Climate Action, a singular focus on the integration of CCE, and in particular climate science, into formal education curricula fails to recognise the significance of supporting the professional development of teachers and the need to address wider structural and contextual constraints that limit more progressive forms of CCE within formal education (NCCA 2015).

Formal education has a crucial role to play in addressing the climate crisis and, increasingly, state policy in Ireland recognises this role. However, policy remains underdeveloped and characterised by the technical focus that underpins wider educational policymaking in Ireland. Current policy responses overly rely on the integration of climate change into curricula, with little consideration for how that curriculum can or will be enacted in schools and classrooms. The next section considers teachers' role in this enactment, the demands on their professionalism and the wider context which shapes their educational practice.

## Emerging Issues for Teachers

CCE places significant demands on teachers, requiring that they are competent in their understanding of climate science, skilful in their navigation of the emotional dimensions of CCE and confident in their capacities to engage with the contested nature of climate justice and climate action within their classrooms. At the heart of this response is the idea of professional agency, that is, the capacity of practitioners to act in particular situations, making sense of policies and of the multiple contextual factors that influence the process by which these policies are implemented. Agency is not a fixed capacity but rather an achievement resulting from the interplay of individual efforts and capabilities within contextual and structural factors in concrete situations (Biesta et al. 2015). Thus, exploring professional agency is key to understanding how educational policies are translated into contextually relevant teaching practices (Martínez Sainz 2018) or, within the context of this chapter, how CCE policies become CCE practices.

Teachers are located within the schools and communities they serve and, given the lack of consensus on appropriate climate action, they may be practising in opposition to the perspectives of other community members. Studies have pointed towards teachers' reluctance to engage with issues that could be seen as controversial, fearing, among other things, a backlash from parents and charges of political indoctrination (Waldron et al. 2016). The controversy outlined in Case Study 1 suggests that such fears have some basis in reality. Despite the compelling evidence that a significant reduction in meat and dairy consumption is required in order to meet climate targets (e.g. Poore and Nemecek 2018), accusations of politicisation, or the promotion of unhealthy lifestyles, contribute to toxifying the context within which teachers make decisions about their CCE practice. Teachers' apprehension towards engaging with controversial issues may also centre on the complexity of such matters and the perceived appropriateness of challenging themes, particularly for younger children. As an existential threat stemming from an issue of complex science, climate change may represent difficult, discomforting knowledge for young learners. However, research has evidenced how learning about global issues, such as climate change, is possible in an early childhood context (Oberman et al. 2014). Barriers to educational practices, such as transformative models of CCE, often relate to teachers' personal understandings of climate change, the narrowness of their conceptualisations of CCE and their beliefs in what constitutes legitimate climate action for young people. In an Irish study, teachers and student teachers viewed climate change as primarily a geographical process, attending predominantly to local manifestations and showing limited awareness of global impact. While some teachers showed concern for intergenerational justice and the need to help vulnerable communities adapt, they displayed limited understanding of causality and of the depth of impact. Participants also focused on private individual actions as the most appropriate response (Waldron et al. 2016). Research elsewhere supports the idea that teachers struggle to consider beyond small, private and individualised actions, such as recycling, responsible consumption or diet (Aarnio-Linnanvuori 2019).

While teachers' lack of competence and confidence is perceived to limit their engagement with global issues (e.g. Bryan and Bracken 2011), teacher education in Ireland has made significant strides in supporting teachers to develop their knowledge and understanding of sustainability issues (Liddy 2012) and the importance of robust CCE within teacher education is recognised (Mochizuki and Bryan 2015; Waldron et al. 2016). Increasing opportunities for professional learning have been found to support teachers to develop their classroom practice for sustainability (e.g. Tarozzi and Mallon 2019), with CCE programmes recognised as part of both initial and in-service teacher education provision in Ireland. Other support for those teachers seeking to address CCE within primary and post-primary schools includes educational resources. While textbooks are recognised as a popular form of educational resource for teachers addressing global issues, variance in content is recognised (Bryan and Bracken 2011), and measured criticism has been levelled at the content of some textbooks. After intervention from An Taisce, for example, a Folens' publication, 'Unlocking Geography', was revised to provide a more accurate representation of the scientific consensus behind anthropogenic climate change (RTÉ 2016). Educational resources which support teachers to develop their pedagogical approaches to climate change have been developed in Ireland and elsewhere, providing frameworks for addressing climate justice and climate action and helping teachers' to develop their own understanding of the field (see, e.g. Creating Futures (Oberman 2016); Sustainability Frontiers (2019); Climate Action: Teacher Resource (Ryan 2018); Eco-Detectives (Pike 2011)). Such publications are themselves open to wider scrutiny (see Case Study 1) and whilst these publications offer support for teachers grappling with CCE, it is also recognised that one particular aspect of CCE remains underexplored—climate action (Selby 2017a).

## Challenges Faced by Children and Young People

Evidence suggests that, as a result of climate change, a reversal in the global improvements in child health and mortality can be expected in a short time (Philipsborn and Chan 2018). Faced with the existential threat of climate breakdown, children and young people from the Global South and across the planet have increasingly been recognised as actors within contemporary debates on climate change action. Media coverage of climate strikes has provided a rare public insight into the concerns of young people across the globe. The climate crisis has also permeated more formal spaces where the views of young people are shared. For example, Dáil na nÓg, the Irish national parliament for young people, identified climate change as one of the most important issues faced by the youth sector in Ireland (Comhairle na nÓg 2019).

Faced with an uncertain and threatening future, the potential for children to experience emotional distress or eco-anxiety has been identified (Pihkala 2017). Supporting children to explore such emotions is a critical part of CCE. Affifi and Christie (2018) argue that CCE needs to incorporate the reality of unsustainability and address the idea of loss, proposing a pedagogy that engages with death as a natural part of the natural world. The overwhelming nature of climate breakdown can provoke a sense of helplessness and lack of agency (Waldron et al. 2016) and educational approaches which are grounded in action have been suggested as a means by which young people may explore collective climate action without becoming lost in the complexity of an issue which can be perceived as insurmountable (Hicks 2014).

Increasingly, academic research has focused on the views of children faced with these significant global challenges. Studies focused on educational interventions have begun to unearth the perspectives of children on their roles and responsibilities in relation to addressing complex issues such as climate change (e.g. Ojala 2012; Martínez Sainz et al. 2019). Given the capacity of younger children to engage with difficult global questions (Oberman et al. 2014), the voices of children within such research provide important insight. Laura, a 12-year-old participant in a research project exploring the efficacy of a CCE resource, explains what she understands of the causes and consequences of climate change but also her desire for action:

> Climate change is when the world is getting hotter, this is because all the carbon dioxide from vehicles and factories is polluting the air. This affects lots of people, animals and countries. Like bees, they pollinate the flowers and we could not live without them. I am trying to do more stuff to help and the people of the world should too. (Martínez Sainz et al. 2019)

While such comments cannot be generalised to all children, Laura's quote expresses her knowledge of the basics of climate science, her understanding of the social and environmental consequences of climate breakdown and her stated commitment to climate action. Alongside the voices of other young people, it highlights the significant shift in the exposure that climate change has received in young people's worlds. It also raises the question of how young people's climate action is understood—what should the people of the world, including children, do about climate change?

In an Irish context, and as illustrated in Case Study 2, while much of the public commentary on the climate strikes was supportive, questions were raised about children's capacity to take independent political action (Gleeson 2019). For the most part, these questions came from outside of education, where ideas of children as social actors have increasingly taken hold. Supported by changing conceptions of children and childhood, which sees children as active agents in the construction of their worlds (e.g. James and James 2012), and by the participation rights enshrined in the United Nations Convention on the Rights of the Child, the idea of children's agency, as noted earlier, has been integral to the development of educational policy and curricula in Ireland since at least the 1990s. While the charge of tokenism has been levelled at some of the consultative and representational structures that have emerged from these developments (McLoughlin 2004; Fleming 2017), recent shifts in policy and practice relating to young people have been more promising, such as the prioritisation of 'listening to and involving children and young people' as one of six national policy goals (DCYA 2014). Earlier critiques of children's participation in schools, which saw it as corralled within safe inauthentic spaces (Waldron 2004), has given way, in some cases, to the visible and public support by schools and teachers of the right to protest. The idea, then, of children's capacities to engage in collective action, arising from their knowledge of the threat posed by the climate crisis and their perception of where the responsibility lies, is one that fits comfortably with current educational perspectives.

## Conclusion

This chapter charted the evolution of CCE towards a multi-dimensional model which integrates climate science, global justice and collective citizenship action, addressing personal, social and environmental concerns, and which necessitates child-centred engagement with the concepts of uncertainty, ambiguity and hope. The chapter has offered an overview of some of the key challenges faced by young people growing up in an era of climate crisis, as they grapple with the complexity of climate science, whilst exploring the collective responsibility to act in light of the immediacy and injustice of climate breakdown. Climate change also presents significant challenges to teachers across the spectrum of experience. Whilst the opportunities to deepen professional understanding of climate change may be present, we recognise that teachers' professional agency is shaped by their surrounding environments and that teachers must be supported to develop strong foundations for engagement with the necessarily interdisciplinary questions of climate action and justice in varied contexts.

When positioned as a part of coherent and coordinated governmental policy, educational policy addressing CCE, in a manner which does not privilege 'soft' apolitical approaches (Andreotti 2006; Selby 2017b), can offer a valuable framework for educators to support the knowledge, understanding and collective action of children, young people and wider society. Ultimately, there is a need to consider our priorities in education, to prioritise the deeper capabilities, values and dispositions that are critical to children's current and future capacities to meet the political, social, emotional and ideological challenges ahead, such as the capacity to move beyond self-interest; to look beyond ourselves, individually and nationally and act in solidarity with people who are not like us; to recognise our shared humanity and believe in the equal worth and dignity of all people—in the words of Martha Nussbaum:

> the capacity to see themselves as not simply citizens of some local region or group, but also, and above all, as human beings bound to all other human beings by ties of recognition and concern. (Nussbaum 2002: 38)

## Case Studies

**Case Study 1: Climate Action Teacher Resource—Education or Indoctrination?**

In April 2019, media reports in Ireland carried a story that the Irish Farmers' Association (IFA) had accused the government of distributing 'propaganda' to schools (Donnelly 2019; Pollak and Hutton 2019) through the launch of a resource for teachers that included references to '#MeatlessMondays' and how to 'reduce the volume of meat and dairy consumed in school' (Ryan 2018: 34). The resource was designed to support teachers to integrate CCE across the junior cycle curriculum and included a welcome letter from the Minister of Communications, Climate Action and the Environment (Ryan 2018). One article carried the headline 'Farmers fume over advice to reduce dairy and meat consumption' and reported:

> An educational guide issued to schools recommending a reduction in meat and dairy consumption has prompted an angry response from the farming sector demanding its immediate withdrawal … Its president Joe Healy dismissed the segment as "propaganda" and said dietary advice offered by An Taisce "crossed the line". (Hillard 2019)

Deemed as pushing an anti-farmer agenda, some activities in the resource were characterised as direct indoctrination of children into a vegan lifestyle (Donnelly 2019). In response, the Minister for Climate Action defended the resource, stating that its intended purpose was to stimulate discussion, reflection and debate. A spokesperson from the Department of Education and Skills distanced the department from the resource and stated that it was down to individual schools to select the materials they use to deliver the curriculum (*Irish Times* 2019).

**Case Study 2: Children and the Climate Strikes**

In Autumn of 2019, thousands of children from over 150 countries became involved in ongoing collective activism demanding government action against accelerating climate breakdown. These events received significant media attention, with considerable focus falling onto Greta Thunberg, a Swedish student who had begun protesting outside of the Swedish parliament. In her speech to the UN Climate Action Summit in New York on Monday, 23 November, Thunberg addressed the audience:

> People are suffering. People are dying. Entire ecosystems are collapsing. We are in the beginning of a mass extinction. And all you can talk about is money and fairytales of eternal economic growth. How dare you!

Despite considerable support for her actions, Thunberg was also exposed to criticism from those questioning a young persons' understanding of and perceived complicity in climate change. Bryan (2019) located the root of this criticism within ageism and sexism. Bryan suggests that this response stems from the challenge to self-interest that climate action represents and the discomforting questions that young people such as Greta Thunberg raise about individual and collective complicity in the breakdown of the earth's climate.

## Bibliography


Aarnio-Linnanvuori, E. 2019. How Do Teachers Perceive Environmental Responsibility? *Environmental Education Research* 25 (1): 46–61.

Affifi, R., and B. Christie. 2018. Facing loss: pedagogy of death. *Environmental Education Research*. https://doi.org/10.1080/13504622.2018.1446511

Andreotti, V. 2006. Soft Versus Critical Global Citizenship Education. *Policy & Practice: A Development Education Review* 3 (Autumn): 40–51.

Ball, S.J. 2003. The Teacher's Soul and the Terrors of Performativity. *Journal of Education Policy* 18 (2): 215–228.

Biesta, G., M. Priestley, and S. Robinson. 2015. *Teacher Agency: An Ecological Approach*. London: Bloomsbury.

Brownlee, M.T.J., R.B. Powell, and J.C. Hallo. 2013. A Review of the Foundational Processes that Influence Beliefs in Climate Change: Opportunities for Environmental Education Research. *Environmental Education Research* 19 (1): 1–20. https://doi.org/10.1080/13504622.2012.683389.

Bryan, A. 2019. *'The Greta Effect'*, RTÉ, 16th September, 2019. Available at: https://www.rte.ie/brainstorm/2019/0916/1076157-the-greta-effect/. Accessed 30 Oct 2019.

Bryan, A., and M. Bracken. 2011. *Learning to Read the World?: Teaching and Learning about Global Citizenship and International Development in Post-Primary Schools.* Dublin: University College Dublin.

Comhairle na nÓg. 2019. *Dail na nÓg 2019.* Available at: https://www.comhairlenanog.ie/dail-na-nog/dail-na-nog-2019/. Accessed 30 Oct 2019.

Department for Communications, Climate Action and the Environment. 2019. *Climate Action Plan 2019: To tackle climate breakdown.* Dublin: Department for Communications, Climate Action and the Environment.

Department for Education and Skills. 2014. *Education for Sustainability: The National Strategy on Education for Sustainable Development.* Dublin: Department for Education and Skills.

Department of Children and Youth Affairs. 2014. *Better Outcomes, Brighter Futures: National Policy Framework for Children and Young People 2014–2020.* Available at: https://www.gov.ie/en/publication/775847-better-outcomes-brighter-futures/. Accessed 30 Oct 2019.

Department of the Environment, Community and Local Government. 2012. *Our Sustainable Future – A Framework for Sustainable Development in Ireland.* Dublin: Department of the Environment, Community and Local Government.

Dijkstra, E.M., and M.J. Goedhart. 2012. Development and Validation of the ACSI: Measuring Students' Science Attitudes, Pro-Environmental Behaviour, Climate Change Attitudes and Knowledge. *Environmental Education Research* 18 (6): 733–749. https://doi.org/10.1080/13504622.2012.662213.

Donnelly, M. (2019). Farmers Anger Over 'Propaganda' as School Children Told to Eat Less Meat and Dairy. *Irish Independent,* April 5. Available at: https://www.independent.ie/business/farming/news/farming-news/farmers-anger-over-propaganda-as-schoolchildren-told-to-eat-less-meat-and-dairy-37985280.html. Accessed 30 Oct 2019.

Fleming, D. 2017. Student Voice: An Emerging Discourse in Irish Education Policy. *International Electronic Journal of Elementary Education* 8 (2): 223–242.

Gaynor, N. 2013. Round Pegs in Square Holes? Development Education, the Formal Sector and the Global Knowledge Economy. *Policy and Practice: A Development Education Review* 23 (Autumn): 1–15.

Gleeson, C. (2019). Dublin School Warns Pupils Off 'Infuriating' Climate Strikes. *The Irish Times,* September 18. Available at: https://www.irishtimes.com/

news/education/dublin-school-warns-pupils-off-infuriating-climate-strikes-1.4022682. Accessed 30 Oct 2019.

Hicks, D. 2014. *Educating for Hope in Troubled Times: Climate Change and the Transition to a Post-Carbon Future*. London: Institute of Education Press.

Hillard, M. 2019. Farmers Fume Over Advice to Reduce Dairy and Meat Consumption. *The Irish Times*, April 4. Available at: https://www.irishtimes.com/news/ireland/irish-news/farmers-fume-over-advice-to-reduce-dairy-and-meat-consumption-1.3849878. Accessed 30 Oct 2019.

Incropera, F.P. 2015. *Climate Change: A Wicked Problem: Complexity and Uncertainty at the Intersection of Science, Economics, Politics, and Human Behavior*. Cambridge, UK: Cambridge University Press.

Irish Times. 2019. Climate Change Strike: Irish Students Join Millions Protesting Globally. *The Irish Times*, September 20. Available at: https://www.irishtimes.com/news/world/asia-pacific/climate-change-strike-irish-students-join-millions-protesting-globally-1.4024673. Accessed 30 Oct 2019.

James, A., and A. James. 2012. *Key Concepts in Childhood Studies*. London: Sage.

Joint Oireachtas Committee on Climate Action. 2019. Report of the Joint Committee on Climate Action Climate Change: A Cross-Party Consensus for Action. Available from: https://data.oireachtas.ie/ie/oireachtas/committee/dail/32/joint_committee_on_climate_action/reports/2019/2019-03-28_report-climate-change-a-cross-party-consensus-for-action_en.pdf. Accessed 31 Jan 20.

Lehtonen, A., A.O. Salonen, and H. Cantell. 2019. Climate Change Education: A New Approach for a World of Wicked Problems. In *Sustainability, Human Well-Being, and the Future of Education*, ed. J.W. Cook, 339–374. Cham: Palgrave Macmillan.

Liddy, M. 2012. From Marginality to the Mainstream: Learning from Action Research for Sustainable Development. *Irish Educational Studies* 31 (2): 139–155. https://doi.org/10.1080/03323315.2011.634099.

Lynch, K., B. Grummell, and D. Devine. 2012. *New Managerialism in Education: Commercialization, Carelessness and Gender*. New York, Houndmills: Palgrave Macmillan.

Martínez Sainz, G. 2018. Building Professional Agency in Human Rights Education: Translating Policy into Practice. *International Journal of Human Rights Education* 2 (1): 1–30.

Martínez Sainz, G., R. Oberman, and B Mallon. 2019. A Framework for Assessing the Impact of Climate Change Education in Schools. ECER Conference. Hamburg, 4 September.

McLoughlin, O. 2004. Citizen Child—The Experience of a Student Council in a Primary School. In *Primary Voices—Equality, Diversity and Childhood in Irish Primary Schools*, ed. J. Deegan, D. Devine, and A. Lodge. Dublin: IPA.

Mochizuki, Y., and A. Bryan. 2015. Climate Change Education in the Context of Education for Sustainable Development: Rationale and Principles. *Journal of Education for Sustainable Development* 9 (1): 4–26.

NCCA. 2009. *Aistear: The Early Childhood Curriculum Framework.* Dublin: NCCA.

———. 2015. *Framework for Junior Cycle.* Dublin: NCCA.

———. 2018. *Sustainable Development: A Study of Opportunities and Linkages in the Primary and Post-Primary Curriculum.* Dublin: NCCA.

———. 2019. *Senior Cycle Review: Consultation Document.* Dublin: NCCA.

Nussbaum, M. 2002. Education for Citizenship in an Era of Global Connection. *Studies in Philosophy and Education* 21 (4–5): 289–303.

O' Sullivan, D. 1989. The Ideational Base of Irish Educational Policy. In *Irish Educational Policy: Process & Substance*, ed. D. Mulcahy and D. O' Sullivan, 219–274. Dublin: IPA.

Oberman, R. 2016. *Creating Futures.* Dublin: Trócaire and the Centre for Human Rights and Citizenship Education.

Oberman, R., F. O'Shea, B. Hickey, and C. Joyce. 2014. *Children's Global Thinking.* Dublin: Education for a Just World.

OECD. 2018. *Preparing our youth for an inclusive and sustainable world: The OECD PISA global competence framework.* Paris. Available at: https://www.oecd.org/education/Global-competency-for-an-inclusive-world.pdf. Accessed 30 Oct 2019.

Ojala, M. 2012. Regulating Worry, Promoting Hope: How Do Children, Adolescents, and Young Adults Cope with Climate Change? *International Journal of Environmental and Science Education* 7 (4): 537–561.

Philipsborn, R.P., and K. Chan. 2018. Climate Change and Global Child Health. *Pediatrics* 141 (6). https://doi.org/10.1542/peds.2017-3774

Pihkala, P. 2017. Environmental Education after Sustainability: Hope in the Midst of Tragedy. *Global Discourse* 7 (1): 109–127.

Pike, S. 2011. *Eco-Detectives Teachers' Resource Pack: Environmental and Climate Change Investigations for Primary Schools.* Dublin: Department of the Environment, Heritage, Local Government and the Centre for Human Rights and Citizenship Education.

Pollak, S., and B. Hutton (2019). IFA Reaction to School's Guide 'Hysterical', says an Taisce', *The Irish Times*, April 5. Available at: https://www.irishtimes.com/news/education/ifa-reaction-to-schools-guide-hysterical-says-an-taisce-1.3851061. Accessed 30 Oct 2019.

Poore, J., and T. Nemecek. 2018. Reducing Food's Environmental Impacts through Producers and Consumers. *Science* 360 (6392): 987–992.

Rousell, D., and Cutter-Mackenzie-Knowles, A. (2019). A Systematic Review of Climate Change Education: Giving Children and Young People a 'Voice' and a

'Hand' in Redressing Climate Change. Children's Geographies. https://doi.org/10.1080/14733285.2019.1614532.

RTÉ. 2016. Folens Revises Chapter on Global Warming. *RTÉ*, 31st August. Available at: https://www.rte.ie/news/2016/0831/813281-an-taisce-global-warming-primary-school/. Accessed 31 Oct 2019.

Ryan, G. 2018. *Climate Action: Teacher Resource*. Dublin: An Taisce Environmental Education Unit.

Selby, D. 2017a. Creating Futures: 10 Lessons Inspiring Inquiry, Creativity & Cooperation in Response to Climate Change for Senior Primary Classrooms. *Policy & Practice: A Development Education Review* 25 (Autumn): 153–160.

———. 2017b. Education for Sustainable Development, Nature and Vernacular Learning. *CEPS Journal* 7 (1): 9–27.

Selby, D., and F. Kagawa. 2010. Runaway Climate Change as Challenge to the 'Closing Circle' of Education for Sustainable Development. *Journal of Education for Sustainable Development* 4 (1): 37–50.

———. 2012. *Disaster Risk Reduction in School Curricula: Case Studies from Thirty Countries*. UNICEF. Available at: https://www.unicef.org/environment/files/DRRinCurricula-Mapping30countriesFINAL.pdf. Accessed 30 Oct 2019.

———. 2015. Development Education and Education for Sustainable Development: Are they Striking a Faustian Bargain. In *Policy and Practice: A Development Education Review*, ed. S. McCloskey, 110–124. Belfast: Centre for Global Education.

Stevenson, R.B., J. Nicholls, and H. Whitehouse. (2017). What Is Climate Change Education? *Curriculum Perspectives*. https://doi.org/10.1007/s41297-017-0015-9.

Sustainability Frontiers. (2019). *Learning Activities*. Available at: http://www.sustainabilityfrontiers.org/index.php?page=learning-activities. Accessed 30 Oct 2019.

Tarozzi, M., and B. Mallon. 2019. Educating Teachers Towards Global Citizenship: A Comparative Study in Four European Countries. *London Review of Education* 17 (2): 112–125.

Thunberg, G. 2019. Think We Should Be At School? Today's Climate Strike is the Biggest Lesson of All. *The Irish Times*, March 15. Available at: https://www.theguardian.com/commentisfree/2019/mar/15/school-climate-strike-greta-thunberg. Accessed 30 Oct 2019.

UNESCO. 2009. *Bonn Declaration on Education for Sustainable Development*. Bonn. Available at: Retrieved from https://unesdoc.unesco.org/ark:/48223/pf0000185055. Accessed 30 Oct 2019.

———. 2015. *Global Citizenship Education Topics and Learning Objectives*. Paris. Available at: http://unesdoc.unesco.org/images/0023/002329/232993e.pdf. Accessed 30 Oct 2019.

———. 2016.*Global Education Monitoring Report: Planet – Education for Environmental Sustainability and Green Growth*. Available at: https://unesdoc.unesco.org/ark:/48223/pf0000246429. Accessed 31 Jan 2020.
———. 2017. *UNESCO Global Action Programme on Education for Sustainable Development: Information folder, 28*. Available at: https://en.unesco.org/themes/119915/publications/all?page=2. Accessed 30 Oct 2019.
Waldron, F. 2004. Making the Irish: Identity and Citizenship in the Primary Curriculum. In *Curriculum and Ideology: Irish Experiences, International Perspectives*, ed. C. Sugrue and F. Waldron, 209–230. Dublin: The Liffey Press.
Waldron, F., B. Ruane, and R. Oberman. 2014. Practice as Prize: Citizenship Education in two Primary Classrooms in Ireland. *Journal of Social Science Education* 13 (1). https:\\doi.org/10.2390/jsse.
Waldron, F., B. Ruane, R. Oberman, and S. Morris. 2016. Geographical Process or Global Injustice? Contrasting Educational Perspectives on Climate Change. *Environmental Education Research* 25 (6): 895–911.

# Climate Action via Just Transitions Across the Island of Ireland: Labour, Land and the Low-Carbon Transition

*Sinéad Mercier, Patrick Bresnihan, Damian McIlroy, and John Barry*

## Introduction

In 2019, the parliaments in Dublin and London declared 'Climate and Ecological Emergencies', as did cities such as Belfast, Dublin and Leeds along with other councils on the island of Ireland from Newry, Mourne and Down to Wicklow. And, in January 2020, Belfast established the

S. Mercier (✉)
National Economic and Social Council, Dublin, Ireland
e-mail: Sinead.Mercier@nesc.ie

P. Bresnihan
Maynooth University, Maynooth, Ireland
e-mail: patrick.bresnihan@mu.ie

D. McIlroy • J. Barry
Queen's University Belfast, Belfast, UK
e-mail: dmcilroy08@qub.ac.uk; j.barry@qub.ac.uk

 
D. Robbins et al. (eds.), *Ireland and the Climate Crisis*, Palgrave Studies in Media and Environmental Communication,
https://doi.org/10.1007/978-3-030-47587-1_14

island of Ireland's first Climate Commission. What are we to make of these declarations and official Commission? On the one hand, these declarations are to be welcomed as belated acceptance of the most recent climate science which has stressed the scale and urgency of reducing greenhouse gas (GHG) emissions. In July 2017, the Republic of Ireland passed a law to ban onshore fracking. However, just days later, one of Ireland's most prominent oil and gas exploration companies, Providence Resources, was granted a license to drill in search of an estimated 5 billion barrels of oil. A year later, the Irish government committed to divest from fossil fuel corporations, the first country in the world to do so. It appears that Ireland's work on climate change is a matter of double-speak—with government taking positive steps, while allowing privatised carbon-heavy industries to undermine them.

## The Origins and Definition of the Just Transition

The origins of the Just Transition are in trade union campaigns to protect workers and communities during the environmental and social damage of the industrial revolution, securing health and safety at work, freedom from disease such as 'miner's black lung' and better living and environmental conditions for the working class. The phrase 'Just Transition' itself was coined in the US trade union movement by Tony Mazzocchi, leader of the Oil, Chemical and Atomic Workers Union, who worked to bring trade unionists into the 'ban-the-bomb' peace movement, alongside a campaign to protect atomic workers in the transition to disarmament (Roessler 2016: 6). Like climate action itself, it is fair to say the Just Transition debate is uneven within public and policy discourse around climate, ecological and energy transition.

In Ireland, both the Irish state and trade union movement have signed up to (if not completely implementing or understanding) the 'Silesia Declaration' (announced at the Conference of the Parties (COP) 24 UN climate conference in 2018) and the relatively under-acknowledged inclusion of 'Just Transition' in the Preamble of the Paris 2015 Agreement. The Silesia Declaration is a commitment by states to tackle climate breakdown recognising that this

> requires a paradigm shift towards building…climate resilient economies and societies…while ensuring a Just Transition of the workforce that creates decent work and quality jobs…and recognises the importance of a

> participatory and representative process of social dialogue involving all social partners. (UNFCC 2018: 1–2)

In November 2019, the Irish Government established the Just Transition Commissioner, and the Green Party put forward their Just Transition Commission (Worker and Environmental Rights) Bill to oversee the implementation of the European Union's (EU) new Governance of the Energy Union package and Ireland's National Energy and Climate Plan. The aim of this bill was to ensure that climate action is taken in a manner that is just and fair to workers, local communities and farmers. Such developments could improve the chances of a Just Transition enabling workers and communities, and not just corporations, to benefit from the green energy transition. And it could also extend a just worker and community-based energy transition to include 'energy democracy' and the democratisation of the ownership and control of renewable energy sources.

At the same time, media discussions of climate action and energy transitions often have little or no consideration as to the social and justice impacts of policy choices. Hence, a benefit of community-based energy transition approaches in Ireland—North and South—is overcoming one of the main obstacles to scaling up renewable energy sources (particularly wind energy): local opposition and lack of social acceptance (see also Chapter "Community Engagement and Community Energy" by Watson, this volume). State-backed provision for community and/or state-owned energy would go a long way towards overcoming that opposition and lay down the parameters for more inclusive and equitable climate and energy action. This would also address the issue of the dominant energy transition strategy being pursued by both jurisdictions on the island, namely, a market-based, neoliberal one. That is, a vision of the low-carbon transition that prioritises and assumes private, corporate ownership of renewable energy production and a heavy focus on 'competitive energy markets'. A key element of any effective Just Transition, therefore, is the 'return of the state', as it were, as a coordinating body to 'steer and manage' any low-carbon energy transition.

Energy transitions are medium-term, (sometimes decades) long processes that need to be strategically managed, mapped and, ideally, based on clarity, inclusiveness, openness and transparency. The shift from one energy system to another is not as simple as switching from one fuel or source to another. One has to include the associated infrastructure that is an integral element of any energy system, from the centralised power stations and long electricity lines to the petrol stations that are part and parcel of the current fossil fuel energy system. As research on energy transitions demonstrates, the transition to a low-carbon, renewable energy economy requires management by the state and a process of social dialogue with affected workers, unions, communities and other stakeholders. This is important not only for ensuring a Just Transition, but also for creating a shared vision amongst all the various energy actors, identifying the most effective policy mechanisms and, finally, for mapping the energy system as a whole, including those energy actors who might not be supportive of any transition to a low-carbon economy.

A Just Transition, a position on the transition to a low-carbon economy endorsed by trade unions, environmentalists, faith communities and major energy actors and policymakers, is about ensuring environmental sustainability as well as decent work, social inclusion and poverty eradication. Most recently, the Paris Climate Agreement of 2015 took into account 'the imperatives of a Just Transition of the workforce and the creation of decent work and quality jobs in accordance with nationally defined development priorities'. However, the fact that it was in the Preamble and not in the main body of the agreement does signal its less-than-central place in that agreement, and it remains to be seen whether subsequent developments such as the Silesia Declaration means that a Just Transition is now a non-optional element of addressing climate breakdown.

As an Irish EPA report on the low-carbon energy transition in Ireland put it, 'governing transitions specifically relates to how to most effectively use the agency of the state and other institutions to steer sociotechnical transformation in desirable directions' (Ellis et al. 2019: 24–25). But beyond that 'functional' role of the state, without state involvement, non-technical issues such as equity and justice are unlikely to figure. Thus, while one could say that the transition to a low-carbon future is inevitable (and we can see movements, uneven as they are, both globally and nationally moving in this direction), whether that transition is just is not. Here we can point to at least two aspects of a Just Transition: on the one hand, if we can see significant socio-economic and other injustices and inequalities within the current economic-energy order, how would simply 'greening/decarbonising business as usual' address those structural injustices (Healy and Barry 2017)? On the other, and this is where the Just Transition discourse connects to other emerging strategic frames and political economy policy platforms such as the 'Green New Deal', the transformative potentials of decarbonising the energy system could at the same time create a different political economy and economy *beyond carbon and beyond capitalism*? (Barry 2012) These are the big questions raised by a Just Transition approach to climate action.

### *The Just Transition Debate on the Island of Ireland*

The Just Transition perspective holds that safe, secure and well-paid green jobs require social dialogue and an active state managing the transition process so that the lowering of GHG emissions does not lower employment or the pay and conditions of affected workers and communities. Here a trade union-centred understanding of Just Transition (which, as noted above, has its origins in the labour movement) stresses that the bargaining power through unionisation is a vital component to achieving those ends. While dominant discourses of the climate-energy transition in Europe and Ireland point to the employment gains brought about by this 'new green industrial revolution', talk and evidence is insufficient. On the one hand, unions have concerns around the potential for these green jobs to be lower paid and precarious. Hence, the consistent call from trade unions for 'decent and well paid jobs'. On the other hand, we sadly have examples, outlined below, of 'Unjust Transitions' on the island of Ireland. And further than that, there is the beginning of a trade union transition discourse around energy as a public good, something that should be thought of a *right* and not a *commodity* (Mercier and Barry 2018). Here, some unions link anti-privatisation campaigns, such as the 'Right2Water' mobilisation which took place across Ireland in 2016, to the low-carbon energy transition. IMPACT trade union stresses the issue of the ownership, control and governance of renewable energy, seeing in a Just Transition the 'opportunity to give citizens a greater stake in low-carbon development through much greater levels of local authority and community ownership of future solar PV, wind farm, biomass and waste-to-energy developments' (IMPACT 2017: 6). But, as outlined below, key here is the role of the state in regulating private energy actors to ensure just and not simply low-carbon energy outcomes, and within the context of the neoliberal character and recent history of the state in both jurisdictions, the struggle for a Just Transition is centrally around a struggle for the role of the state in managing and coordinating any low-carbon energy transition.

A salient point here is that unlike other countries, Ireland, while heavily dependent on imported carbon energy, does not have a large carbon extraction and processing industry. However, it does have many workers in the carbon energy and electricity production sector, much of which is heavily unionised. For Ireland, North and South, as outlined below, the agricultural sector is the equivalent of the carbon energy extraction sector in other European countries such as Spain and Germany, given the agricultural sector's economic importance and contribution to greenhouse gas emissions. It is also an industry that does not provide a liveable income for most farmers. The decarbonisation of agriculture and how this sector can be transformed to reduce its GHG emissions is still a major missing piece from the dominant discourse of a Just Transition in terms of a new industrial revolution. This is especially true in relation to how government plans for the expansion of the sector (the Food Wise 2025 plan) run completely counter to its climate change commitments.

Another, as yet unacknowledged benefit of a greater *indigenisation* of the Just Transition concept is the integration of the Northern Ireland peace process within localised understandings of how a Just Transition should be implemented. The peace process, imperfect as it is, does have lessons around conflict management and transformation that should be useful in activism and strategy around Just Transitions. This is because any transition, even a just one, will produce 'winners' and 'losers' and will face opposition and inertia from dominant interests. Conflict is inevitable, and it is better to prepare for it than be taken by surprise. Here, learning and amending some of the lessons from conflict transformation processes such as the one in Northern Ireland could be important additions to the governance of any Just Transition process. The requirement for widespread worker and community involvement in the 'social dialogue' around the content of any energy or sustainability transition, and thus the possibility of serious differences and conflicting views, could benefit from the experience of the Northern Ireland peace process. And this could be, alongside the island's abundant renewable energy sources, the island of Ireland's distinctive contribution to the global debate and localised struggles for a Just Transition.

The resurrection of the NI Executive and Assembly in January 2020 illustrates that the Northern Ireland peace process (so far) remains resilient, even in the mouth of Brexit. Whatever happens with direction of travel for Brexit it is assumed that the governance structures and institutions that were created within strands one, two and three of the Good Friday Agreement will endure. Nevertheless, the people of Ireland north and south are now faced with two major issues: delivery on the obligations of the 1998 Good Friday Agreement (GFA) and tackling climate breakdown. The Good Friday Agreement never considered the local implications of global climate breakdown; it is a peace agreement constituted to deliver a stable post-conflict society in Northern Ireland and manage complex political relationships.

However, the recent *New Decade, New Approach* document, which outlines the Assembly's Programme for Government, has a dedicated section on climate change which sets out a raft of new commitments, including an energy strategy that references Just Transition, the development of a Climate Change Act and the establishment of an independent Environmental Protection Agency (House of Commons Library 2020: 44). The inclusion of this section within the Programme for Government is significant, not only because it outlines clear commitments to tackle the climate emergency, but also because it creates potential points of parity and contact for an island-wide approach. The *environment* remains an institutionalised area of North/South cooperation under strand two of the GFA and as such has the potential to reinvigorate broader peacebuilding efforts on the island of Ireland through collectively negotiating the island-wide priorities to tackle the climate emergency. Just Transition as a social, economic and environmental necessity moves beyond borders and thus creates the potential to act as a conflict resolution and peacebuilding mechanism that could and should be led by the interregional Trade Union Movement with deep and historical organisational structures across the island.

### *Examples of Unjust Energy Transitions*

While climate change (alongside energy security concerns) requires that we have to retire the fossil fuel energy system, this cannot happen overnight. It requires planning and long-term thinking, not short-term, market-based decisions made solely or largely around costs. Mitigating measures need to be put in place to ensure a smooth and a Just Transition from a high- to a low-carbon economy. As the energy transition should not disproportionately negatively impact communities and workers dependent on oil, coal and gas energy production, nor should it disproportionately impact citizens who are carbon energy consumers (especially those who are in energy/fuel poverty). It is clear that any transition to a low-carbon economy requires that the fossil fuel energy system declines over time and then is replaced by a renewable energy system. However, we have examples of badly structured and managed energy transitions within both jurisdictions in Ireland, which does not bode well for inclusive and just energy transition and climate action. In Tullamore, County Offaly, we have the Bord na Móna decision to close a peat-fired power plant, while in Kilroot, County Antrim, we see the closure of a coal-fired plant (Barry 2018).

The decision to close two Electricity Supply Board (ESB) peat-firing power stations—in Lanesboro, County Longford, and in Shannonbridge, County Offaly—will have a devastating impact on the midlands and the workers, communities and others dependent on those stations for employment and livelihoods. As an Irish Congress of Trades Union report on the decision put it, 'unfortunately, what was also revealed in the wake of the announcement was the absence of a plan for a Just Transition with respect to the Bord na Móna workforce, and the glaring absence of such an overall national strategy that would help ensure that workers and local communities across the country are not simply abandoned to their fate' (ICTU 2019b: 5). The lack of consultation and meaningful engagement by Bord na Móna with local trade unions, workers and communities in this case can be viewed as an example of how ***not*** to organise a Just Transition—a situation that the Irish state can ameliorate by learning from the ongoing German Coal-Exit Commission (Agora Energiewende 2019) or the process which led to all privately owned coal mines being closed in Spain in 2019. The company failed to abide by either the Silesia declaration principles or the International Labor Organization guidelines, which demand that any Just Transition policy framework should 'anticipate impacts on employment, adequate and sustainable social protection for job losses and displacement, skills development and ***social dialogue, including the effective exercise of the right to organise and bargain collectively***' (ILO 2015: 6; emphasis added). What has happened with the closure of Kilroot power station near Larne in County Antrim is the premature retirement of a piece of the fossil (in this case coal) energy mix in Northern Ireland, and like in the Bord na Móna case, without consultation with workers and other affected stakeholders or structural adjustment packages to assist in the building of a new green and indigenous industry with high-quality jobs (ICTU 2019a; Barry 2018).

## A Just Transition for Agriculture

The concept of a Just Transition most often applies to workers in the energy sector. This is understandable: the move to a low-carbon future will most directly affect workers involved in the extraction and processing of fossil fuels. In Ireland, however, this situation is complicated by the fact that the greatest source of greenhouse gas emissions is the agricultural sector. This raises different questions and challenges for a Just Transition on the island of Ireland. GHG emissions from the agricultural sector account for over 30% of the total emissions in Ireland—the highest proportion from any agricultural sector in the European Union (EPA 2019a). This is due to Ireland's concentration on dairy and, to a lesser extent, beef farming. These sectors are also the main pressure on water quality. The EPA's most recent national report on water quality found that there has been a continuous decline in surface water quality since 2012 largely due to nutrient run-off from agriculture (EPA 2019b). As well as emissions and water quality, the agricultural sector is the main driver of habitat decline according to the National Parks and Wildlife Service (NPWS). In a recent report (2019), they identified that agricultural practices are negatively impacting over 70% of terrestrial and coastal habitats.

Beginning in 2019, protests over beef prices, including blockades of Dublin city centre by incensed beef farmers, have surfaced the chronic problems around farm viability in a context of cheap food and inequitable global food supply chains. TASC's recently published report on inequality in Ireland highlights that the agricultural sector has the most severe inequality in income distribution of any sector in Ireland (TASC 2018). Another, less discussed source of inequality is the still dominant culture of male-only primogeniture. A Macra na Feirme study in 2014 revealed that only 11% of farming successors (to farmers over 50 years) are female (Macra na Feirme 2014: 8). This has led to Ireland having by far the EU's lowest number of women working in the agricultural sector (Moran 2018). This places women 'on the perimeter of mainstream farming in Ireland' (Mulhall and Bogue 2013: 3) with negative results in terms of farm sustainability. Gender equality is necessary for a sustainable rural economy (Shortall and Byrne 2009); therefore, rural development and agricultural policies in a Just Transition require a holistic view of community well-being that incorporates a gender analysis.

Brexit will undoubtedly have a negative impact on the economic fortunes of the agricultural sector. This is particularly the case for farmers in the North of Ireland. With so many farm households, and rural communities, reliant on the single farm payment under Common Agricultural Policy (CAP), it is unclear how, or even if, these payments will continue to be made once Britain leaves the EU. The EU-Mercosur deal (agreed in principle in 2019) will further undermine the viability of beef farming in

Ireland, an already struggling sector that also faces changing consumer habits and the likelihood of increasing extreme weather events such as those that hit the farming sector in 2017 and 2018.

While today's social, economic, (geo)political and environmental uncertainties are unprecedented, it is important to situate the current moment within the longer 'transition' that Irish agriculture has undergone over the past 50 years. If this history is ignored then there is a danger that the term Just Transition will become another consensus buzzword, conditioned and applied in the same way as environmental initiatives and rural diversification schemes have been since the 1990s. There is a body of critical literature from rural sociology and geography that responded to the re-structuring of Irish agriculture, and rural Ireland more generally, in the crucial period of the 1980s/1990s. The continued relevance and value of this scholarship means it is worth returning to. At the same time, the relative dearth of similar academic scholarship in the intervening years reflects a shift towards urban-focused research within social sciences, something that needs to change if critical scholars hope to shape the direction of the Just Transition process here. By critically examining why efforts to resolve the structural contradictions of the Irish agricultural sector have failed, we can begin to see where energies and attentions need to be channelled in order to bring about more meaningful, and just, transitions within the sector.

In Ireland, entry into the European Economic Community (EEC) in 1973 opened up new opportunities and avenues for farmers to expand their productivity, particularly in the dairy sector. As well as EU subsidies, technological developments (including major infrastructures such as networked water supplies and rural electrification) and, latterly, improved technical expertise relating to cattle breeds, chemical fertilisers and pesticides, combined to transform farming practices and rural landscapes. In 1988, Teagasc (the state-supported institute for agricultural research and extension) was set up to further improve efficiency on farms, shifting focus away from output and towards financial and business management. While Teagasc was nominally focussed on productivity and efficiency in 'the agricultural sector' as a whole, the need to raise national levels of output became a justification for concentrating programmes of aid for technological investment on the most 'progressive' farmers, who were already those with the most resources (Tovey 2000).

What took shape from the 1980s was a dualistic agricultural system in which one class of (dairy) farmers were encouraged to become more efficient enterprises and another class of farmers were forced to diversify into other sectors, including the less profitable beef sector. As early as 1996, large specialist dairy farms had an average income of £58,000, while small farms in dry cattle production had an average income of £1800. The main 'successful' policies, from the point of view of the majority of farmers who were small scale, thus came from government departments that had nothing to do with agriculture, namely, rural industrialisation and social welfare policies. It is this process of stratification within the farming sector that has continued apace since the 1990s and which provides the context within which Just Transition debates need to take place.

Since the 1980s, the production of milk has been subordinated to the needs of an expanding global food industry (Tovey 1991). Dairy cooperatives, which have been consolidating since the 1960s, have also sought to respond to a situation of growing competitiveness in European and world markets that was greatly facilitated by an era of free trade in agricultural commodities. Rather than food production, farming was being re-framed as the production of inputs for the food industry as dairy cooperatives moved into processing and the provision of ingredients to the food industry. All of which increased both the carbon energy intensity of the sector and its GHG emission profile.

Probably far more significant than employment potential, in terms of influencing state policy, are the opportunities for accumulation generated in the food industry, which are not available in farming (Tovey 2000). In 1995, the cooperative food industry (primarily dairy processors) recorded a turnover of £6000 million, compared to £20 million in 1970. Since the 1990s, the global expansion of the dairy processing cooperatives has only accelerated, with the opening of Chinese markets and diversification into niche ingredient supply chains. It could be argued that Ireland's 'agricultural' economy is more accurately part of Ireland's 'supply chain' economy, where processing, business management, finance and marketing take precedence over food production and where the profits accrue to these downstream activities. And yet, when responsibility is assigned for environmental problems related to the sector, it is farmers that are called out, not the CEO of Glanbia or the marketing director of Kerry Group on six figure salaries.

The continued rationalisation of agriculture in Ireland and across the EU has resulted in a trend towards 'fewer farms, less farm employment, larger farm units and the specialisation and concentration of farm production' (Ní Laoire 2002: 16). The commercialisation of Irish agriculture has also shifted social norms and institutions in Irish farming society, from 'family farming' to 'one-man-farms', mutual aid to isolation and marginalisation, and a lowering in the national status of farmers and values (Macken-Walsh 2010). In 2011, 70% of farms were non-commercially viable and survived only through the off-farm work of female spouses (Macken-Walsh 2010), leading Derwin to conclude that 'in a peculiar sense, it is women's off-farm work that keeps farming male' (Shortall 2006: 313).

The question of a Just Transition for the farming sector is part of a bigger set of questions relating to rural Ireland and the stalled or failed project of Irish 'agrarian modernity' (Tovey 2000). Other directions were possible: Michael Davitt's demands to socialise land, articulating Ireland's struggle against Landlordism as part of a global, anti-imperial struggle of indigenous, colonised peoples; the cooperative movement of the same era, which offered a different kind of agrarian modernity, one which sought to embrace opportunities of modern technology but through egalitarian control and ownership of these productive forces. These roads not taken can still provide inspiration, re-centring the need for a more egalitarian land ownership system—particularly the removal of the damaging culture of male-line primogeniture and the need to create a more cooperative farming system with younger family members while the landowner is living, rather than waiting for a will.

A Just Transition in Ireland has to think big and grasp the reality that cheap food will not last forever and cannot continue to be outsourced to other parts of the world. It is worth remembering that extreme, climate-related events don't just affect agricultural production here, they also expose the fragility of Ireland's food supply chains; the images of empty food shelves after Storm Emma hit Ireland in 2018 may have been a novelty then, but they should also be seen as a sign that the current global agri-food system is not sustainable and needs to be replaced with diverse alternatives that connect producers and consumers in more egalitarian and ecologically just relations. A Just Transition for the sector needs to imagine an agricultural sector that can provide meaningful land-based work *and* safe, nutritious food through biodiverse, low-carbon farming. This challenges the dominant discourse within EU and national policy that imagines 'surplus' farmers transitioning into providers of environmental or touristic services to urban dwellers. Linking Just Transition and the need for alternative food (and land-use) systems also challenges the media-fuelled narrative of rural versus urban Ireland and farmers versus environmentalists. Examples of these alternatives already exist and are growing (the land-workers network Talmamh Beo in the Republic, for example). The problem is that they remain marginal to the dominant model advanced by state policy and agencies. A Just Transition for the agricultural sector thus requires a well-funded Teagasc supporting and training young farmers in mixed farming and agro-ecology, rather than continuing to focus on grass and milk yields.

## Labour, the Environment and a Just Transition

Services Industrial Professional and Technical Union (SIPTU) organiser for ESB and Just Transition campaigner, Adrian Kane (2018) has stated that 'I think, if we are to be honest, like many others we have been late in the day in seriously engaging with climate change in the Trade Union Movement'. This section presents the argument that an alternative and intersectional framing of what is considered 'environmentalism' in Ireland reveals that trade unions have a far longer history of involvement than presented, giving some hope for the Just Transition idea (with its origins in the trade union movement) to act as a coalition-builder between labour and the climate and ecological movement in Ireland. The call for a Just Transition is not only a continuation of that trade union involvement, but also links to a human rights turn in climate/energy politics (see above on campaigns to view energy as a 'right' not a 'commodity'). It is also an expression of what has always been an Irish popular environmentalism from 'below', prioritising the concerns of community organisations, rural identity, Irish trade unionism and farming against exploitation.

### *History of Irish Environmentalism in the South*

Hilary Tovey (1993) and later Liam Leonard (2007) trace the history of environmentalism in Ireland as a conflict between 'a growth-based form of ecological modernization' driven by 'official' environmentalism, versus a grassroots localised 'rural populism', based around a reflexive construction of local rural identity which expanded as Ireland became more urbanised. 'Official environmentalism' is the dominant discourse of environmentalism in Ireland, historically preoccupied with the maintenance of 'urban and rural "heritage" in the form of visual amenities, historical residues and ecological complexes' (Tovey 1993: 413; Garavan 2007). Mainstream conservation and preservations strategies in Ireland tend to adopt a variant of Northern European environmentalism that sees environmental damage as 'what have we done to ourselves' (Tovey 2007: 21).

Some environmental historians, echoing similar critiques by activists in the Global South (Guha 1999), have criticised what they see as the Malthusian underpinnings of Northern European environmentalism, which tends to promote a lens that an overpopulating and ignorant poor are the primary causes of environmental damage (Björn-Ola 2003). This leads to the orientation of governance away from what is assumed to be an ignorant 'collective' towards ensuring the adoption of 'vetted' techniques and ways of seeing that prioritise a technocratic 'expert class' that can wield the language of state and corporate power and work within the policymaking process. In climate campaigns, this means a narrow focus on lobbying and legislation to ensure technocratic solutions are adopted in government policy, often market-based solutions such as emissions trading schemes and carbon pricing (Doherty and Doyle 2013: 131, 180). In the 2019 Report of the Joint Oireachtas Committee on Climate Action (JOCCA 2019), one of the most prominent and controversial recommendations, an increase in carbon taxation, could be viewed as an example of this 'market-based' and 'technocratic' climate action.

Despite the high media and public profile for national 'official' environmental organisations, much of the actual environmental mobilisation that occurred from the mid-1980s to the mid-1990s appears to have happened independently of national-level organisations (Tovey 1993). It was located particularly in rural areas of the country under threat from inappropriate forms of 'development' and 'modernisation'. These localities could be said to be suffering from 'ecological marginalisation' and were also (in the later cases of fracking and local opposition to wind energy developments) self-perceived as 'sacrifice zones' for both carbon and renewable energy-based forms of industrialisation and modernisation. This rural and place specificity of localised opposition around energy transitions is important in understanding 'unjust energy transitions' such as the Bord na Móna case in the Irish midlands discussed briefly above.

The rapid growth of industrialisation in the wake of the Celtic Tiger boom brought benefits that largely bypassed rural areas, which became repositories for multinational investment in mining, chemical factories, industrial production processes and others. Opposition campaigns were launched by 'local activist groups who did not see themselves as engaged in "environmental" mobilisation and drew help and support from networks of "community" or trade union activists in other locations' (Tovey

2007: 27). This dichotomy continued into the early 2000s as professional non-governmental organisations (NGOs) focused on European and global campaigns on climate legislation and individual carbon footprints (Doherty and Doyle 2013: 180–2), as well as 'climate justice' which highlighted the low contribution to climate breakdown of people in the Global South. Ironically, such campaigns occurred as local groups and trade unions fought in the Corrib conflict alongside the Nigerian Ogoni people against Shell, now known to be one of the top 100 corporations that have collectively contributed to 71% of global $CO_2$ emissions since 1988 and has been a major contributor to systemic climate denial (Jacques et al. 2008).

### *An All-Ireland Environmentalism: The Just Transition*

The Irish trade union movement has been involved in environmental activism for decades (Milton 2003) and has a language that shares elements with Irish popular (and rural) environmentalism. In her study of the Irish environmental movement, Tovey (2007: 21) notes that the trajectory of Irish environmentalism has more in common with the Global South and southern European environmentalism with experience of autocratic rule (ibid.: 25), which have a local-level rather than national-level focus. These local discourses of climate action or environmentalism are framed in terms of 'the disorganisation of biological processes, the locals' loss of their resource base, and the generation of a wide range of socio-economic, political and public health risks' (Tovey 2007: 26). This framing of postcolonial independence also operates in opposition to environmental regulation 'from above'—as locals across the island of Ireland rejected Dublin-based, EU or state regulations such as bog-cutting regulations created without their involvement as 'neo-colonial governance'—echoes of which can be found in the Bord na Móna case above with rural communities feeling 'left out, and left behind' by decisions made in Dublin. Another example is the Irish water protests, defended in a framework of 'polluter pays' by some environmental NGOs as against a framing of anti-austerity, extractivism and a human right to water by predominantly working class, trade union and rural campaigners (Hearne 2015).

Trade unions in Ireland share a common set of values, origins and increased shared objectives with environmental and climate justice movements. Both have origins in protecting the public good, including public health and the built urban environment in the case of the green movement, and both share an aim to improve the welfare and living conditions of people, place and, now with climate breakdown, planet. With this framing, a link can be made to Latin American environmental campaigns against 'extractivism' in Ireland. During the 1980s and 1990s in South and Central America, neoliberal policies re-focused development agendas towards extracting resources, in tandem with a rollback on workers' rights and public services, leading to high murder rates of land protectors and trade unionists. In fighting such extractivist policies, local groups work with other organisations facing oppression, including trade unions. Climate NGO campaigns in the Global North are slowly shifting from the technocratic rhetoric of carbon footprints, pricing and trading towards an 'anti-extractivist' focus (Klein 2016; Stewart 2015). This is due to the influence of the disruptive rhetoric of indigenous and anti-racism groups, youth climate strikers, Extinction Rebellion, 350.org and the Carbon Majors report and court cases which focus on the corporate producers and sponsors of coordinated climate denialism campaigns, such as Exxon Mobil, or the Dutch Urgenda case (see chapter "Climate Litigation, Politics and Policy Change: Lessons from Urgenda and Climate Case Ireland" by O'Neill and Albas, this volume). Such an analysis dovetails well with trade unionism, which historically has called for collective organising and state action in order to combat the exploitation of communities and the environment by powerful (often transnational) corporations.

## Conclusion

The Kilroot and Tullamore decisions throw into sharp relief the need to have longer-term plans for the transition of the island's energy system and economy. And this should be a political decision, led and coordinated by the state, with mandated democratic decision-making processes to include communities and workers, not simply left to market actors and processes. Energy and electricity in particular are strategic and vital resources to the economy and society that require political and long-term management. If not directly nationalising energy, a 'just energy transition' to a low-carbon economy requires more direct democratic state 'governance' of that process, with a long-term transition plan to ensure people, communities and workers are not unfairly disadvantaged. If a transition away from coal, oil and gas means job losses, economic disadvantage and threats to the economic security of communities, with no mitigating measures or transition plan for those communities in place, then people will not unreasonably be sceptical or outright resist such proposals. Both these cases could be said to make public support for low-carbon energy transitions much more difficult. Rather than being part of a managed and planned process of shifting our energy system away from fossil fuels, these decisions may succeed in bolstering those (including within the trade union movement) who wish to continue our carbon-based energy system and delay or dilute climate action.

As we move forward in the creation of national and international campaigns for climate action hopefully framed within a Just Transition, trade unions bring to the climate movement a valuable class and political understating and toolkit of collective class consciousness and organising against an exploitative system, that is, capitalism. However, trade unionism must be updated and move past growth-based productivism and its anthropocentric framing and add a 'post-carbon' agenda, as well as revive its more long-standing critique of capitalism. The integration of justice concerns for land and labour, for people and place, is essential (if not easy to coordinate) for climate action and energy transitions in Ireland, if those actions are to be actively supported by citizens as both jurisdictions on the island of Ireland speed up their overdue transition to a low-carbon and climate-resilient economy and society.

## Bibliography


Agora Energiewende. 2019. *The German Coal Commission: A Roadmap for a Just Transition from Coal to Renewables.* Available at: https://www.agora-energiewende.de/fileadmin2/Projekte/2019/Kohlekommission_Ergebnisse/168_Kohlekommission_EN.pdf. Accessed 31 Oct 2019.

Barry, J. 2012. Climate Change, "The Cancer Stage of Capitalism" and the Return of Limits to Growth: Towards a Political Economy of Sustainability. In *Climate Change and the Crisis of Capitalism*, ed. M. Pelling, D. Manuel-Navarrete, and M. Redclift, 129–143. London: Routledge.

———. 2018. Closure of Kilroot – Or How Not to Manage a 'Just Transition' to a Low Carbon Energy Economy. *QPOL.* Available at: http://qpol.qub.ac.uk/kilroot-closure-transition-low-carbon-energy/. Accessed 31 Jan 2019.

Björn-Ola, L. 2003. *The Return of Malthus: Environmentalism and Post-War Population-Resource Crises.* Cambridge: White Horse Press.

Breathnach, P. 2000. The Evolution of the Spatial Structure of the Irish Dairy Processing Industry. *Irish Geography* 33 (2): 166–184.

Doherty, B., and T. Doyle. 2013. *Environmentalism, Resistance and Solidarity: The Politics of Friends of the Earth International.* London: Palgrave Macmillan..

Ellis, G., T. Hume, J. Barry and R. Curry. 2019. *Catalysing and Characterising Transition*. Available at: http://www.epa.ie/pubs/reports/research/climate/Research_Report_287.pdf, Accessed 10 Jan 2019.

EPA. 2019a. *Water Quality in Ireland, 2013–2018*. Dublin: EPA.

———. 2019b. *Ireland's Greenhouse Gas Emissions Projections 2018–2040*. Dublin: EPA.

Fórsa. 2019. *Fórsa Backs Climate Demonstrations*. Available at: https://www.forsa.ie/forsa-backs-climate-demonstrations/. Accessed 10 Jan 2019.

Garavan, M. 2007. Resisting the Costs of 'Development': Local Environmental Activism in Ireland. *Environmental Politics* 16 (5): 844–863.

Guha, R. 1999. *Environmentalism: A Global History*. London: Pearson.

Healy, N., and J. Barry. 2017. Politicizing Energy Justice and Energy System Transitions: Fossil Fuel Divestment and a "Just Transition". *Energy Policy* 108: 451–459.

Hearne, R. 2015. The Irish Water War. *Interface: A Journal for and About Social Movements* 7 (1): 309–321.

House of Commons Library. 2020. *New Decade, New Approach*. Available at: https://commonslibrary.parliament.uk/parliament-and-elections/devolution/northern-ireland-assembly-new-decade-new-approach/. Accessed 31 Jan 2020.

Impact. 2017. A Just Transition to a Low-Carbon Economy Implications for IMPACT and its Members. Available at: https://www.iiea.com/ftp/Publications/2017/IMPACT%20JustTransition.pdf. Accessed 20 Feb 2020.

International Labor Organization. 2015. *Guidelines for a Just Transition towards Environmentally Sustainable Economies and Societies for all*. Geneva: ILO.

Irish Congress of Trade Unions. 2019a. *Building a Just Transition: The Case of Bord Na Mona*. Available at: https://www.ictu.ie/download/pdf/building_a_just_transition_report_feb_2019.pdf. Accessed 20 Jan 2020.

———. 2019b. Trade Unions and Climate Campaigners Call for Immediate Establishment of National Just Transition Taskforce. 19 September. Available at: https://www.ictu.ie/press/2019/09/19/trade-unions-and-climate-campaigners-call-for-imme/. Accessed 20 Jan 2020.

Jacques, P., R. Dunlap, and M. Freeman. 2008. The Organisation of Denial: Conservative Think Tanks and Environmental Scepticism. *Environmental Politics* 17 (3): 349–385.

Joint Oireachtas Committee on Climate Action. 2019. *Report of the Joint Committee on Climate Action, Climate Change: A Cross-Party Consensus for Action*. Available at: https://data.oireachtas.ie/ie/oireachtas/committee/dail/32/joint_committee_on_climate_action/reports/2019/2019-03-28_report-climate-change-a-cross-party-consensus-for-action_en.pdf. Accessed 29 Jan 2020.

Kane, A. 2018. Speech to ICTU Energy & Natural Resources Conference on, ICTU Just Transition Conference, May 11. Available at: https://www.ictu.ie/globalsolidarity/newsevents/keyissues/2017/05/11/just-transition-conference/. Accessed 10 Dec 2019.

Klein, N. 2016. Let Them Drown: The Violence of Othering in a Warming World. *London Review of Books* 38 (11). Available at: https://www.lrb.co.uk/the-paper/v38/n11/naomi-klein/let-them-drown. Accessed 10 Jan 2020.

Leonard, L. 2007. Environmentalism in Ireland: Ecological Modernisation Versus Populist Rural Sentiment. *Environmental Values* 16 (4): 463–483.

Macken-Walsh, A. (2010). A Review of the Social Benefits of Joint Farming Ventures. *The Rural Economy Research Centre Working Paper Series: Rural Economy Research Centre*. Teagasc: Athenry.

Macra na Feirme. 2014. Land Mobility and Succession in Ireland. Available at https://www.ifa.ie/wp-content/uploads/2014/01/Land-Mobility-and-Succession-in-Ireland-Report.pdf. Accessed 9 Jan 2020.

Mercier, S., and J. Barry. 2018. Progress towards a Just Transition on the Island of Ireland. United Nations Research Institute for Social Development. *Just Transition Forum*. Available at: https://medium.com/just-transitions/barry-mercier-89d92079701f. Accessed 31 Jan 2020.

Milton, K. 2003. *Environmentalism: The View from Anthropology*. London: Routledge.

Moran, C. 2018. New Figures Show Ireland Has by Far the Least Female Farmers in Europe. *Irish Independent*, 3rd January. Available at: https://www.independent.ie/business/farming/agri-business/new-figures-show-ireland-has-by-far-the-least-female-farmers-in-europe-36451722.html. Accessed 9 Jan 2020.

Mulhall, L., and P. Bogue. 2013. Women in Agriculture. *National Rural Network*. December.

Ní Laoire, C. 2002. Young Farmers, Masculinities and Change in Rural Ireland. *Irish Geography* 35 (1): 16–27.

NPWS. 2019. *The Status of EU Protected Habitats and Species in Ireland 2019*. Volume 1: Summary Report. Dublin: NPWS.

Roessler, C. 2016. "Just Transition" – Just What Is It? An Analysis of Language, Strategies, and Projects, Labor4Sustainability. Available at: https://www.labor4sustainability.org/files/Just_Transition_Just_What_Is_It.pdf. Accessed 3 Dec 2019.

Shortall, S. 2006. Economic Status and Gender Roles. In *Rural Gender Relations: Issues and Case Studies*, ed. B.B. Bock and S. Shortall, 303–317. Wallingford: CABI Publishing.

Shortall, S., and A. Byrne. 2009. Gender and Sustainability in Rural Ireland. In *A Living Countryside – The Politics of Sustainable Rural Development*, ed. J. McDonagh, S. Shorhall, and T. Varley. Aldershot: Ashgate Publishing.

Stewart, L. 2015. James Orr: Environment Is the Forgotten Victim of Northern Ireland Peace Process. *Belfast Telegraph*, 5 January. Available at: https://www.belfasttelegraph.co.uk/news/environment/james-orr-environment-is-the-forgotten-victim-of-northern-ireland-peace-process-30881356.html. Accessed 3 Dec 2019.

TASC. 2018. *Cherishing all Equally 2019: Inequality in Europe and Ireland.* Dublin: TASC.

Tovey, H. 1991. Of Cabbages and Kings: Restructuring in the Irish Food Industry. *Economic and Social Review* 22 (4): 333–350.

———. 1993. Environmentalism in Ireland: Two Versions of Development and Modernity. *International Sociology* 8 (4): 413–430.

———. 2000. *Milk and Modernity: Dairying in Contemporary Ireland. Research in Rural Sociology and Development.* Bingley: Emerald Group Publishing.

UNFCC. 2018. Solidarity and Just Transition Silesia Declaration. Available at https://cop24.gov.pl/fileadmin/user_upload/Solidarity_and_Just_Transition_Silesia_Declaration_2_.pdf. Accessed 31 Jan 2020.

# The Changing Faces of the Climate Movement in Ireland

*Lorna Gold*

## Introduction

The literature on the global climate movement points to the diversity that exists within civil society action on global warming. Sociological analysis can help to understand the emergence of this movement, and how it has behaved at key junctures in history, such as the Copenhagen Summit in 2009. Understanding the power dynamics which shape social movements, how issues are framed and the different roles played by different actors is helpful in learning lessons in the struggle against climate breakdown. This is particularly true at times of rapid change, where different frames and actors are challenged—but some more than others. The Irish climate change movement has gone through rapid growth and transformation since 2015. In this chapter, I explore the new contours of the Irish climate movement within the context of the global climate movement and consider some of the key strategic and organisational challenges it faces at a time of rapid evolution.

L. Gold (✉)
Department of Applied Social Studies, Maynooth University, Maynooth, Ireland
e-mail: Lorna.Gold@trocaire.org



D. Robbins et al. (eds.), *Ireland and the Climate Crisis*, Palgrave Studies in Media and Environmental Communication,
https://doi.org/10.1007/978-3-030-47587-1_15

## Civil Society, Activism and Climate Justice

The Irish climate movement is part of the much bigger story of the global climate movement, which is itself embedded within the contested spaces of global civil society. The idea of a 'global civil society', whilst contested, is generally viewed as the place 'outside' the market and the state defined by voluntary association amongst citizens (Schaefer et al. 2013). The realm of civil society is where citizens meet, discuss, form associations and mobilise on issues of concern. Civil society is composed of a wide variety of actors with different objectives, aims and strategies. Amongst these, non-governmental organisations (NGOs) have historically played a central role as 'a broad spectrum of formal non-profit organisations that are independent of government, do not advocate violence and seek to advance public goods' (Betsill 2015: 252).

According to social movement theorists (Edwards and McCarthy 2001), civil society becomes social movement when groups of citizens start to frame their shared grievances together and gain the capacity to mount large-scale campaigns to address those grievances. This capacity involves both human and financial resources which enable collective action to emerge and be sustained over time. Social movements are often linked to major existing infrastructure within civil society, such as faith institutions, universities or trade unions. The existence of social infrastructure through which a movement can organise provides scale and reach and lowers the transaction costs in the establishment of new campaigns. When this capacity is combined with political spaces to make change possible, shared strategies and tactics can emerge; a social movement is born.

The global climate movement emerged in the 1990s as an off-shoot of the larger environmental movement which came to the fore in the 1970s and 1980s. The focal point for the global climate movement over the past 30 years has largely been the formal United Nations Framework Convention on Climate Change (UNFCCC) process which emerged out of the UN Rio Earth Summit in 1992. Environmental NGOs were key to the emergence of this global UN process in the 1980s through their work to highlight the environmental implications of unchecked economic growth, epitomised in the World Commission on Sustainable Development Report *Our Common Future*, published in 1987 (World Commission on Sustainable Development 1987). This engagement has been documented by scholars since the mid-1990s as part of the literature on global environmental politics after the Rio Summit in 1992 (Newell 2006). This literature has focused on the role of NGOs in shaping ideas and norms in international politics and the role of NGOs in raising awareness about the climate problem as activists.

The build-up to and failure of the Copenhagen Summit in 2009 marked a pivotal moment in the emergence of the climate movement as distinct from the environmental movement (Lipschutz and McKendry 2011). Deep divisions emerged in the climate movement during the Copenhagen summit—at a time of peak growth and rapid evolution. Since 2015, as global emissions have continued to rise, divergent views have emerged within the climate movement as to the continued focus on the UNFCCC process as the primary vehicle for bringing about change. Divergent, translocal climate justice struggles have emerged bringing together development, human rights and climate change (Trócaire 2017). Most recently, the explosion of the climate strike movement, led by activist Greta Thunberg, has ignited new and diverse membership and focus into the climate movement.

### *Defining the Climate Movement*

In 2007, the Intergovernmental Panel on Climate Change (IPCC) put forward three roles played by civil society in the context of the UNFCCC negotiations, which have been the focus for climate advocacy at a global level since their initiation in 1992 at the Rio Earth Summit. According to the IPCC, civil society plays critical roles by providing policy advocacy, policy research and political spaces for new political reforms. Newell (2006: 98–101) proposes a typology of three groupings, clustered around certain roles played by the climate movement in the context of the UNFCCC negotiations. The first are the 'insider-insiders', who focus on 'green governmentality', negotiating on the inside of negotiations on the basis of science evidence. Their aim is primarily to shift the negotiations forward via internal pressure and persuasion. The second grouping are the 'insider-outsiders', who are both active on the inside of negotiations but also work outside via campaign tactics such as petitions, letters and putting pressure on negotiators through the media. Their objectives vary, but generally are focused around the idea of ecological modernisation and robust governance in tackling the climate crisis. The final grouping cited are the 'outside-outsiders' who are primarily focused on putting pressure on negotiations via external pressure. Their objectives also vary, but range from a focus on civic environmentalism to radical resistance to the existing political and corporate power structures. Their goal is to 'challenge[s] the neoliberal approach embedded in ecological modernisation and call[s] for radical democratisation of global governance and economic processes' (Schaefer et al. 2013: 246).

The role of NGOs within this movement is key. Betsill (2015: 251) sees NGOs as playing three overlapping roles—'as activists raising awareness and calling for action; as diplomats working with governments to craft climate policies; and as governors developing new mechanisms for steering society towards a low-carbon future'. NGOs act as activists when they marshal the issue onto the political agenda. The NGO climate activism tends to take place within the spaces of institutionalised climate politics—especially within the UN. Fulfilling this function is highly dependent on NGOs having access to those political forums. NGOs, moreover, also act as 'diplomats', playing similar roles to governmental delegates at international forums—except the NGOs represent a global constituency, not a country. They are generally present as observers, highly organised through constituency groups. Finally, Betsill sees NGOs as acting increasingly as 'global governors' in the climate policy regimes emerging through monitoring functions, shadow reporting and building capacity.

According to Hadden (2015), the roles adopted by those engaged in the collective action of civil society groups are dependent on a number of critical factors. If individuals are working from within established organisational structures, for example large NGOs, they have a 'repertoire' of specific actions they can choose from. This largely frames the choices they make in climate activism. In many ways, these are dictated by the types of organisations they represent and the relationships that exist between them. Organisations reliant on professionals and contractual obligations are largely bound by what is legal, unless they have made explicit provision to support employees in unlawful protests, for example. If someone is employed to do professional policy analysis and advocacy on climate negotiations, an organisation can't expect them to risk arrest through street activism.

A related issue is examined through the question of professionalisation of civil society organisations, especially NGOs. This points to the inadequacy of how individuals and organisations work within networks (Gold 2016). The assumption of 'voluntarism' within civil society (meaning that individuals are engaged in action due to an intrinsic motivation) is somewhat displaced when NGOs become highly professionalised and work largely within market constraints like any corporations. Large, complex

NGOs with reputations, brands and donor funding to protect may be very slow to engage in activities that are contentious and could upset any one of those stakeholders (Gold 2016).

### *Framing the Issue*

The roles played by NGOs and groupings are indicative of the frames various groups use. Schaefer et al. (2013), in detailed empirical research on the climate movement in the USA, point to at least six dominant prognostic frames at play within the climate movement in the run up to the fateful 2009 Copenhagen Summit, where cleavages in the climate movement came to the fore. The dominant frames identified were system change, global justice, individual behaviour change, technology change and investment, policy and legislative change and change to industry and production.

One key factor in the cleavages which emerged during the Copenhagen Summit in 2009 is highlighted by Routledge (2011). He points out that in previous summits, the dominant civil society groupings had primarily centred around a global narrative of emissions reductions. This narrative was framed in terms of science evidence rather than a moral argument. However, in the build up to 2009, the 'climate justice' narrative emerged from the Global South as a counter to this Northern-driven science-based narrative. Routledge describes the climate justice narrative as 'referring to attempts to conceptualise the interrelationships between, and address the roots causes of, the social injustice, ecological destruction and economic domination perpetrated by the underlying logics of capitalism that has seen industrialized countries reap the benefits of fossil fuel intensive development' (Routledge 2011: 385).

As such, the climate justice movement embodied local struggles which were deeply embedded in individual struggles, such as the landless farmers' movement in Bangladesh and La Via Campesina movement. These struggles pre-existed the climate movement, but their struggles are even more urgent given the impact of climate change. The engagement of these actors, and their narrative of injustice, in the broader climate movement exposed new and competing stories. Far from being a 'movement of movements'—in the sense of a mega-movement with singular objectives and plans—the emerging climate justice movement is a collection of 'overlapping, interacting, competing and differentially placed and resourced networks concerned with issues of climate change and justice' (Ibid.:

385). Within these overlapping struggles, there are many opportunities for building 'translocal solidarity'—and also many risks of division.

The global emissions narrative and the spatial embeddedness of the climate justice movement sat uneasily side by side, despite both claiming to share an overarching objective of tackling climate emissions. This work of connecting the spatially embedded, localised struggles to secure rights and claims with the broader shared narrative of climate injustice became the dominant focus of the climate movement in the post-Paris world. It involves recognising the spatial particularisms of how climate change plays out. 'This must negotiate the problem of how militant particularisms concerning climate change transcend local concerns to form common ground upon which solidarities can be constructed' (Ibid p. 386).

The new networks and coalitions that emerge, however, are always subject to pre-existing power patterns. There is a mismatch of access, resources and capacity to act. Whereas in these movements, for large organisations, acting for climate justice may be a programmatic issue—a matter of funding partners—for those in the Global South actively engaged in struggles, it is a matter of life or death. Networks are key to the emerging translocal climate justice struggles, but these are both influenced by and replicate the existing power geometries. This power geometry is reflected in access to resources but also in the asymmetry of impacts different organisations and individuals face in the struggle.

In the current context, however, in the face of the growing youth climate movement, this localised narrative is once again challenged. The youth climate strikers, and the new wave of protest movements, are focusing attention once again on the a-spatial narrative of time and urgency. This narrative, which highlights the shared time deficit to keep global emissions within safe limits for future generations, has the potential to displace the translocal solidarities. To be sustainable in the long run, the new wave of climate activism has to be mindful of the consequential impact of the interconnections between global-local narratives for climate justice.

## The Climate Movement in Ireland

The environmental movement in Ireland has been deeply rooted in local struggles since the 1970s. Leonard's overview of the movement outlines its roots in the anti-nuclear movement in the 1970s and then in numerous local campaigns against aspects of rapid economic development as the 'Celtic Tiger' took hold (Leonard 2008). Local campaigns were often associated with a loss of rural identity and a sense of place. The first civil society coalition specifically on climate change, Stop Climate Chaos (SCC), was established in 2006 as a collaboration between Friends of the Earth Ireland and Trócaire. Over the following decade or so, this coalition expanded to include around 30 organisations, mostly with a strong urban base in Dublin. The coalition membership grew slowly, with a substantial number of civil society organisations supporting the network being from the international development sector and almost none from the environmental movement that pre-existed the coalition. During the 2000s, the Stop Climate Chaos coalition faced a number of challenges, especially under-resourcing, having only ever had a part-time coordinator supporting activities for the whole network. The movement was composed largely of paid employees of the participating civil society organisations.

Despite the small scale of Stop Climate Chaos, the coalition had some key strengths, which enabled it to have a significant impact on the policy environment. Its key strength was in its policy capacity, political access and communications reach. Despite the relatively low levels of public engagement in public events and demonstrations, SCC developed a strong ability to provide robust policy analysis and coordinated lobbying. Leveraging its constituent members' (especially Friends of the Earth and Trócaire) advocacy, SCC played a prominent role in key policy initiatives including the introduction of an Irish climate bill, proposals around a carbon tax, climate finance and the Irish position in relation to the Paris Agreement in 2015. The coalition also contributed to policy debates around key European Union (EU) positions on renewables targets and Irish contributions to EU targets, working as the Irish member of various international climate movement coalitions such as major European networks such as 'Climate Action Network' (CAN) Europe and the Catholic NGO network, CIDSE. During this period, the dominant role played by SCC was that of the 'insider-outsider', providing science-based policy advice and diplomacy, but at the same time seeking to engage the public via stunts and media presence. The group worked as a cohesive block, occupying the political space as the 'civil society voice' in key domestic and international processes virtually uncontested.

As with the global climate movement, the signing of the Paris Agreement in December 2015 as the successor to the Kyoto Protocol marked a key milestone for SCC. In the cycle of key advocacy objectives, it represented a point of arrival and opportunity to evaluate the impact of the Irish climate movement and its future. Whilst SCC had been an active player in key processes in the years running up to the Paris Agreement, very few of its positions were actually reflected in Ireland's official policy positions. Despite nearly a decade of insider advocacy, the Irish government continued to adopt entrenched policies working against climate action. Even the much-lauded climate bill was weak to the point of being ineffectual as a legal instrument. Within the Paris negotiations, the Irish state had advocated and successfully ensured that agricultural interests were embedded in the Paris Agreement through pushing for the inclusion of the clause 'does not threaten food production' in Article 2.1.b. Moreover, key governmental policies, such as the Food Harvest 2020 and Food Wise 2025, actively promoted an increase in dairy and meat production, undermining the potential for meeting Ireland's climate obligations.

In 2015, despite years of climate advocacy, it seemed that the public remained staunchly reticent to any significant measures to address the climate crisis. Public priorities were firmly focused on other issues, dominated in large part by the financial crisis and long recession. Climate change rarely, if ever, came up on the doorstep as a burning political issue. On the contrary, the installation of windfarms, the potential ban on cutting peat and the growing narrative around the need to 'cull the national herd' to reduce emissions positioned the climate movement as anti-rural communities. The dominant base of the coalition in urban, Dublin-based organisations became a significant weakness. A number of rural politicians and media outlets readily jumped on this narrative which pitched the climate movement against the farming community. It was clear that the climate movement in Ireland faced a serious challenge in communicating the science of climate change in ways which resonated and engaged a wide swathe of public opinion.

### *Contours of the Evolving Movement*

It is against this backdrop that the climate movement has been changing in Ireland since 2015, and especially since late 2018. The new wave of the movement did not start in one place or have one particular founder. Rather, it has emerged in many spaces where different groups have found their own way to the climate science for different motivations.

In the aftermath of the Paris Climate Agreement, two new climate campaigns quickly took hold in Ireland almost simultaneously. Both campaigns marked a shift away from targeting official governmental processes towards other climate actors. The first campaign was a campaign to follow in the footsteps of several European countries to ban hydraulic fracking in Ireland (Halloran 2017). This campaign connected the broader issue of climate action with local concerns, particularly in Leitrim, Cavan, Sligo and Fermanagh, around the impacts of fracking on the local environment. The 'Love Leitrim' campaign connected up with national groups such as Friends of the Earth. The second campaign built on the success of the anti-fracking campaign. It was the fossil fuel divestment movement, first spearheaded by campaigner Bill McKibben and taken up in early 2016 by Trócaire and Maynooth University (MU). Trócaire and MU were soon joined by numerous student groups advocating for their own universities to divest from fossil fuels, most notably Trinity Fossil Free Campaign.

The emergence of the fracking and divestment campaigns marked new departures for climate campaigning in Ireland in a number of ways. Firstly, these were some of the first campaigns which did not focus attention on the agricultural sector, which has long been the most intractable issue for Irish climate policy. Instead, they sought to draw attention to the equally challenging issue of our ongoing investment in the future of the oil and gas industry at a time when science is unequivocal about the need to phase out fossil fuels. Alongside the universities, Irish churches also stepped in to make commitments to divest their funds from fossil fuels, pointing to the moral obligation to care for the earth and promote social justice within the Christian faith, as highlighted in *Laudato Si* (Pope Francis 2015).

Furthermore, the 'new politics' that emerged following the General Election in March 2016 presented a major political opportunity for the climate movement. This new political make up resulted in the two dominant parties Fine Gael and Fianna Fáil sharing an uneasy partnership in government, whilst the opposition was composed of an unprecedented number of independent TDs. Within this context, it became possible to move forward with the Citizens' Assembly and the National Dialogue on Climate Change, both of which emerged from ideas circulating within the Green Party and the wider climate movement. Moreover, the new political arithmetic meant that innovative pieces of legislation on fracking and fossil fuel divestment could be introduced. Different groups supported new TDs to bring forward private members bills focused on challenging the fossil fuel industry. Tony McLoughlin, a Fine Gael TD from Sligo, brought forward the anti-fracking bill in early 2016. Working closely with independent Donegal TD Thomas Pringle, a coalition of climate activists supported the passage of the private members' Fossil Fuel Divestment Bill through both houses of the Oireachtas. Despite Fine Gael opposition, the

Fossil Fuel Divestment Bill was passed into law in the summer of 2018, representing the single most important victory for the climate movement in Ireland in recent times. It also represented a sign of hope to the whole global climate movement which was seeking to re-evaluate its future following the Paris Agreement.

These legislative successes gave the Irish climate movement a new buoyancy. However, they also exposed the massive contradictions at the heart of government climate policy and the intransigence in tackling the broader issues of Ireland's ongoing climate inaction. It became evident that the increasingly stark warnings of scientists were not being translated into robust and urgent action. Whilst the government had been forced into some innovative climate action due to the political arithmetic, these did not translate into any change in official predictions of future emissions trajectories. The National Mitigation Plan 2017 (DCCAE 2017), which was a statutory obligation under the 2015 Climate Act, revealed a wish list of uncosted and, at times, spurious proposals to tackle climate change. Examples include Action 5—'Prepare periodic reports on the electricity and gas prices for households and businesses'—and Action 20—'Finalise Wind Energy Guidelines'. Whilst the many reviews, reports and studies outlined in the mitigation plan may be important, none committed to actual emissions reductions. It was this mismatch between the proposed policy framework, the science and the legislation which prompted another small climate action group, 'Friends of the Irish Environment', a coalition member of SCC, to bring forward a legal case in early 2018 to hold the government to account (see chapter "Climate Litigation, Politics and Policy Change: Lessons from Urgenda and Climate Case Ireland" by O Neill and Alblas, this volume).

### *Strikes and Rebellion*

The year 2018 marked a distinctive shift in the climate movement in Ireland and internationally. In late 2018, two international movements, Extinction Rebellion and Fridays for Future, both involving different kinds of non-violent direct action and civil disobedience, emerged in Ireland. Whilst no in-depth academic studies have yet examined the impact of these movements, preliminary research undertaken in autumn 2019 for this chapter showed the extent of the upsurge in climate groups. During 2019, nine new national groupings and 40 new local Irish climate action groups had registered on social media platforms such as Twitter and

Facebook. As these new groups emerge, so do many of the power dynamics discussed above, exposing the diversity of narratives and roles at play.

The Extinction Rebellion movement, which started in the UK in 2018, committed to peaceful non-violent direct action to alert the public to the imminent threat of ecological collapse. The focus of the movement is on the crisis of biodiversity loss and on global heating as one of the main causes, alongside other human activities which lead to a loss of habitat and pollution. Urgency is a key message of all XR actions. The first organised XR activities in the UK were reported in the summer of 2018. In Ireland, the first day of action was in early December 2018 in Dublin and Cork. The organisation has become much more sophisticated, developing its capacity to manage peaceful, yet disruptive, demonstrations. The philosophy underpinning the group is one of proposing widespread Gandhian non-violent direct action—something that has not been seen in the Irish climate movement until now—though this was present within the ecological movement that predated the climate movement (Leonard 2008).

The other movement that took root in Ireland in 2019 was the Climate Strike movement started by Swedish activist, Greta Thunberg. A weekly climate strike was started by a mixed group of concerned parents and students in December 2018 and continued on most Fridays through 2019 outside Leinster House. Linking in with local schools in Dublin, a number of teachers accompanied their students to participate in the weekly protests. This weekly strike movement had the effect of highlighting the urgency of the crisis and the need to start to engage in disruptive action for change.

An upswell of student climate action started to emerge from the start of 2019. A small number of self-organised teenage climate strikers started to skip school on Fridays and follow Greta's lead. The 'Fridays for Future Ireland' and 'School Strike 4 Climate Ireland' groups emerged almost simultaneously, advocating similar disobedience to highlight the climate crisis. These two groups operated on the basis of striking on Fridays—that is, not seeking permission from school and being prepared to face the consequences. As well as these two groups, a separate school-supported student group, 'Schools Climate Action Network' (SCAN), quickly emerged as a key player in the rapidly evolving climate strike movement. This network operated quite differently to the Fridays For Future and School Strike 4 Climate groups in that it sought to engage the school community in the protest movement. Adults, especially teachers, played a key role in SCAN and were absolutely central to ensuring school participation in the major climate marches held in 2019, culminating in tens of thousands striking on 20 September. At the same time, connections were forming with the Irish Congress of Trade Unions, which came on board to support the strikes through logistics. SCAN members, however, played a much smaller role in the regular climate strikes held weekly in various locations across the country.

### *Green Faith Rising*

In the midst of fast evolving Irish climate movement, a more subtle shift has been happening across other parts of civil society in Ireland, most notably in faith communities. The environmental group Eco-Congregation Ireland has been active in Ireland for over a decade, highlighting the connections between Christianity and care for the environment. The group has been engaging parishes in looking at how they can become more ecological, advocating measures to lower emissions and increase biodiversity, as well as incorporating ecological messages into church services. Since 2015, this work has gained new significance following on from Pope Francis's encyclical *Laudato Si – On Care for our Common Home*. The Irish Bishops' Conference established a working group in September 2016 to examine possible measures which the Irish Church could take to implement Laudato Si at a national and diocesan level, including theologians such as Fr. Sean McDonagh and leading climatologist Professor John Sweeney from Maynooth University. Amongst the measures approved have been the decision to divest from fossil fuels, the approval of an annual 'Season of Creation' in September in all Irish parishes, a focus on climate action during Catholic schools week and the integration of climate and care for creation into religious education curriculum at different levels.

During Pope Francis's visit to Ireland in 2018, a special focus was placed on sustainability through the 'Our Common Home' project. Out of that initiative, a national network of Laudato Si animators and parishes was established, with a strong focus on engaging with local groups to build coordinated local climate action. Alongside the diocesan and parish structures, many hundreds of religious orders became active supporters of climate activism in their local areas. Ten Irish religious orders signed up to divestment from fossil fuels between 2018 and 2020, with many more actively considering shifting their resources towards care for creation. The Church of Ireland, Irish Bishops Conference and a majority of Catholic dioceses also supported divestment. Many parish groups actively engaged in and even led the Fridays for Future and Extinction Rebellion protests at a local level, such as the Catholic parish in Dundrum, as well as spearheading educational and awareness raising initiatives in collaboration with other secular groups.

### *Localised Climate Action Groups*

The combination of the climate strikes, Extinction Rebellion and resurgent interest of faith-based groups in climate activism, together with growing media coverage of the issue, has galvanised a rapidly evolving climate movement in rural and urban communities alike. Harnessing the organising power of social media as the main means of communication, local action groups are emerging across the country. These geographically defined groups tend to overlap with the national groups, embedding the national climate action with local context and issues.

## A Rapidly Evolving Movement

As with any emerging and rapidly evolving social movement, the diversity of players and their frames sooner or later comes to the fore (Hadden 2015). The rapid evolution of the Irish climate movement bears some parallels with the phase of growth in the global climate movement that preceded the Copenhagen Summit in 2009. In particular, the rapid growth in the Irish movement throughout 2019 saw the emergence of multiple actors across different sectors. It also saw a shift in tone away from the nuanced technical policy language, towards climate injustice. Tactics started to change too—with the 'insider' groups matched (or surpassed) by multiple 'outsider' groups deploying more direct action. These shifts arguably worked to engage the public on an emotional level and mobilise larger numbers of people than previously seen. However, to become effective change it has to translate into specific measures and policy proposals. While it is very straightforward to be against something—inaction, delays and climate change—it has been less easy to articulate a clear, united voice in favour of action in a dispersed, self-organising movement that aspires to be 'leaderless' (but established leaders already exist!).

This task of articulating political asks of a growing and diverse movement is not straightforward. The political engagement space is already (still) largely occupied by SCC, which now represents a small fraction of the movement as a whole. This pattern reflects the pre-existing power dynamics where in an emerging network, pre-existing dynamics and resources dictate who is heard (Routledge 2011). One clear example is the introduction of a carbon tax. On the one hand, there are some groups which strongly advocate a high carbon tax levy on polluting activities such as flying or meat consumption. On the other hand, others would argue strongly that individuals should not be penalised for their activities when ready alternatives do not exist. In its Budget 2020 submission, Social Justice Ireland pointed out that those with the lowest incomes would be hit the hardest by the carbon tax, particularly those in fuel poverty. The issue of just transition becomes central to any policy platform. In the run up to the 2020 General Election, SCC played a critical role in addressing this challenge by convening a 'connect' meeting to bring all climate groups—local and national—together. The result was an election campaign—One Future 2020—which incorporated a significant number of groups.

As the Friday youth strike movement continues, the ongoing role of schools and teachers is going to come under increasing scrutiny, generating potential divisions between those students calling for more weekly strikes and those who wish the 'climate action' to be incorporated into the school activities. The schools' infrastructure has provided a very useful jumping off point for climate action, but the regular strikers say this weakens the purpose of the strike, arguing that something that is sanctioned by school is not a strike. On the other hand, there is a huge pressure on schools to respond to the demand of students to voice their anxiety about the climate emergency. One way this has been resolved in many schools has been through parent engagement. Students are facilitated to participate in marches without an absence penalty so long as they do so with parental consent and/or accompaniment and catch up on work later. Teachers unions have been supportive of schools taking this line; the Government, however, has remained mute.

## Conclusion

The Irish climate movement is undergoing a time of unprecedented change, though it is too early to put numbers on it. This period of change in some ways mirrors the transition which happened within the climate movement globally a decade ago when the climate justice narrative came to dominate, with the resultant cleavages between activists. As the movement continues to evolve and new groups grow, with overlapping mandates and identities, the literature around the global climate movement and social movements in general can help frame the discussion. In considering the Irish climate movement, it helps to consider it as more diverse than homogenous. Within the movement, there are pre-existing power dynamics, different overlapping roles and resource challenges. In particular, a major question is what happens to the older, more established groups such as Stop Climate Chaos as largely 'insider' players when the demands of the more 'outsider' players grow stronger. In analysing the movement, however, it is important to recognise that whatever typology is used, however you dissect it, the weight of the movement rests principally in its ongoing mobilisation capacity. Engaging in climate activism, in the words of Hadden, is like fish feeding on a coral reef. 'These groups pursue different strategies and demands. The cumulative effect of their advocacy has results above and beyond their individual actions' (Hadden 2015: 168). The important thing is not that everyone is saying the same thing, but that there is advocacy happening—and for that, we need plenty of fish.

## Bibliography


Betsill, M.M. 2008. Environmental NGOs and the Kyoto Protocol Negotiations: 1995–1997. In *NGO Diplomacy: The Influence of Non-governmental Organisations in International Environmental Negotiations*, ed. M.M. Betsill and E. Corell, 177–206. Cambridge: MIT Press.

Betsill, M. 2015. Chapter 22: NGOs. In *Handbook on Climate Governance*, ed. K. Backstrand and E. Lovbrand, 251–261. Cheltenham: Elgar.

Cohen, J.L., and A. Arato. 1992. *Civil Society and Political Theory*. Cambridge: MIT Press.

DCCAE. 2017. *National Mitigation Plan*. Available at: https://www.dccae.gov.ie/en-ie/climate-action/publications/Documents/7/National%20Mitigation%20Plan%202017.pdf

Doyle, T., and S. MacGregor, eds. 2013. *Environmental Movements Around the World: Shades of Green in Politics and Culture*. Santa Barbara: Praeger Press.

Duyck, S., S. Jodoin, and A. Johl. 2015. *Routledge Handbook of Human Rights and Climate Governance*. Abingdon: Routledge.

Edwards, B., and J.D. McCarthy. 2001. Resources and Social Movement Mobilisation. In *The Blackwell Companion to Social Movements*, ed. D.A. Snow, S.A. Soule, and H. Kriesi, 116–152. Oxford: Blackwell.

Gerten, D., and S. Bergmann, eds. 2013. *Religion in Environmental and Climate Change: Suffering, Values, Lifestyles*. London: Continuum International.

Gold, L. 2016. Are International Non-Governmental Development Organisations Fuelling a 'Conspiracy of Silence' Around Climate Injustice? *Development* 59: 243–250.

Grierson, J., and V. Dodd. 2020. Terrorist List That Was Circulated Across Government Included Extinction Rebellion. *The Guardian*, January 27. Available at: https://www.theguardian.com/environment/2020/jan/27/terror-police-list-extinction-rebellion-shared-across-government

Hadden, J. 2015. *Networks in Contention: The Divisive Politics of Climate Change*. Cambridge: Cambridge University Press.

Halloran, M. 2017. Ireland Joins France, Germany and Bulgaria in Banning Fracking – Private Members' Bill Passed in Seanad After Passage Earlier Through Dáil. *Irish Times*, June 28. Available at: https://www.irishtimes.com/news/politics/oireachtas/ireland-joins-france-germany-and-bulgaria-in-banning-fracking-1.3137095

Kearns, L. 2011. The Role of Religions in Activism. In *The Oxford Handbook of Climate Change and Society*, ed. J.S. Dryzek, R.B. Norgaard, and D. Schlosberg. Oxford: Oxford University Press.

Keck, M.E., and K. Sikkink. 1998. *Activists Beyond Borders: Advocacy Networks in International Politics*. Ithaca: Cornell University Press.

Leonard, L. 2008. *The Environmental Movement in Ireland*. Dordrecht: Springer.

Lipschutz, R., and C. McKendry. 2011. Social Movements and Global Civil Society. In *The Oxford Handbook of Climate Change and Society*, ed. J. Dryzek, R. Norgaard, and D. Schlosberg, 369–383. Oxford: Oxford University Press.

Newell, P. 2006. Climate for Change? Civil Society and the Politics of Global Warming. In *Global Civil Society 2005/6*, ed. M. Glasius, M. Kaldor, and H. Anheier, 90–119. Thousand Oaks: Sage.

One Future. 2020. Questions to Ask Candidates. Available at: https://www.onefuture.ie/assets/files/pdf/questions_to_ask_candidates_about_climate_change.pdf

Pope Francis. 2015. *Laudato Si – On Care for our Common Home*. Rome: Vatican Press.

Routledge, P. 2011. Translocal Climate Justice Solidarities. In *The Oxford Handbook of Climate Change and Society*, ed. J.S. Dryzek, R.B. Norgaard, and D. Schlosberg. Oxford: Oxford University Press.

Schaefer, B., R.J. Brulle, and A. Szasz. 2013. Civil Society, Social Movements and Climate Change. In *Climate Change and Society – Sociological Perspectives*, ed. R.E. Dunlap and R.J. Brulle. Oxford: Oxford University Press.

Schäfer, M. 2015. Climate Change and the Media. In *International Encyclopedia of the Social & Behavioral Sciences*, ed. J.D. Wright, 2nd ed., 853–859. Amsterdam: Elsevier.

Scholsberg, D., and L.B. Collins. 2014. From Environmental to Climate Justice: Climate Change and the Discourse of Environmental Justice. *Wiley Interdisciplinary Reviews: Climate Change* 5 (3): 359–374.

Trócaire. 2017. Still Feeling the Heat – How Climate Change Continues to Drive Extreme Weather in the Developing World. Available at: https://www.trocaire.org/sites/default/files/resources/policy/still-feeling-the-heat.pdf

World Commission on Environment and Development. 1987. *Our Common Future Report*. Oxford: Oxford University Press.

# Cloughjordan Ecovillage: Community-Led Transitioning to a Low-Carbon Future

*Peadar Kirby*

## Introduction

As the pressure mounts on governments and on citizens to take more radical and impactful action on climate change, concrete examples of how to undertake such change become ever more important. This chapter examines one such model, Cloughjordan Ecovillage (CEV) in County Tipperary, Ireland. In her book on ecovillages, Karen Litfin calls them 'living laboratories … running collective experiments in every realm of life: building, farming, waste management, decision-making, communication, child rearing, finance, ownership, aging and death'. She sees them as 'opportunities for learning', both for ecovillagers themselves and for the wider society (Litfin 2014: 18). The author of this chapter is a resident of Cloughjordan Ecovillage, and during the writing of this chapter, Litfin's words took on a poignant quality as four members died, two of them very suddenly. As with so much else, this challenged us to grieve well as a community, to find appropriate ways of celebrating those we had lost and to

---

P. Kirby (✉)
University of Limerick, Limerick, Ireland
e-mail: Peadar.Kirby@ul.ie



D. Robbins et al. (eds.), *Ireland and the Climate Crisis*, Palgrave Studies in Media and Environmental Communication,
https://doi.org/10.1007/978-3-030-47587-1_16

begin picking up the pieces. This illustrates how ecovillages seek to engage collectively with the many challenges of life, but framed always by practices of sustainability and transitioning to an ever lower-carbon lifestyle.

The chapter begins by asking what is an ecovillage, drawing on the definitions of the Global Ecovillage Network (GEN). It then introduces Cloughjordan Ecovillage as a project in progress, its founding objectives, and how far it has got towards fulfilling those objectives in its 20 years of existence. The third section surveys current published research on the ecovillage project. The discussion then identifies the three features that define this intentional community as an ecovillage: its ecological building standards, its carbon-neutral district heating system (DHS), and its food system, centred on its community farm, run on the principles of Community Supported Agriculture (CSA). The central importance of community to the ecovillage's understanding of sustainability is next highlighted, introducing its governance structures, and charting the ups and downs of a community weathering many challenges, especially the severe economic downturn of 2008. The chapter finishes by outlining CEV's educational work and by looking to the future of the ecovillage.

## What Is an Ecovillage?

Cloughjordan Ecovillage is a member of the Global Ecovillage Network (GEN) which defines an ecovillage as 'an intentional or traditional community using local participatory processes to holistically integrate ecological, economic, social, and cultural dimensions of sustainability in order to regenerate social and natural environments'. It sees them as one solution to the great problems of our times—the limits to growth and the unsustainability of our societies. It states: 'Ecovillages, by endeavouring for lifestyles which are "successfully continuable into the indefinite future", are living models of sustainability, and examples of how action can be taken immediately. They represent an effective, accessible way to combat the degradation of our social, ecological and spiritual environments. They show us how we can move toward sustainability in the 21st century (Agenda 21)'. In 1998, ecovillages were first officially named among the United Nations' (UN) listing of 100 Best Practices, as excellent models of sustainable living (GEN website).

Reflecting on her time spent in 14 very different ecovillages around the world, Litfin (2014) identifies community life as being the common feature distinguishing all of them—'the intangible kinds of sharing that are the essence of community'. Based on this, she concludes that 'the foundation for ecological sustainability is social sustainability, person to person. In many of the ecovillages I visited, I saw concrete demonstrations that a self-replenishing social order is based on relationships of trust and reciprocity' (Litfin 2014: 16–17). This emphasis is echoed in Jackson and Jackson's reflection on the history of GEN:

> Ecovillages provide models for lifestyles that reduce our ecological footprint while delivering a better quality of life – lifestyles that are possible in all countries of the world. The models are based on solidarity and cooperation and may provide a prototype that can lead to global justice. In ecovillages we are learning how to solve conflicts, how to develop a global consciousness, how to create places where children can grow up in sane and healthy environments, how to use renewable, integrated energy systems, how to provide 100% organic food and how to live lives full of love and compassion. (Jackson and Jackson 2015: 217)

Transitioning to a low-carbon society will also depend on a lot of social innovation, requiring experimentation with new ways of living within our ecological footprint. In a paper for the European Union (EU) on transitional governance in the service of sustainable societies, Belgian political scientist Olivier De Schutter (2014: 17) emphasises the 'role of social innovations empowering people to invent local solutions'. He writes that these social innovations abound and they are often local and territory-based. 'They typically are based on hybrid governance structures, bringing together municipalities, the private sector, the "third sector" and non-governmental organisations or citizens' groups' (ibid.). He gives the example of transition towns and mentions Cloughjordan Ecovillage as an example, describing it as 'a supportive social community living in a low-impact way to create a fresh blueprint for modern sustainable living' (ibid.).

The importance of bottom-up experimentation in creating new approaches towards sustainability has also come to the fore in the 2018 report from the Intergovernmental Panel on Climate Change (IPCC). Pathways towards a low-carbon society, as presented in earlier IPCC reports, have been criticised for their dependence on as yet unproven technologies, particularly BioEnergy with Carbon Capture and Storage (BECCS). This dependence has been described by Anderson and Peters as 'an unjust and high stakes gamble' due to its reliance on 'a highly speculative technology' (quoted in Hickel 2019: 56). However, the 2018 IPCC report on the 1.5 °C target for global warming, in speaking of the scale of transformations required, makes clear that these cannot be limited to a narrow view of decision-making focused on technical solutions because this 'tends to crowd out more participatory processes, obscures contested values and reinforces power asymmetries' (Roy et al. 2018: 459). It states that what it calls 'past development trajectories' do not go far enough because they 'can constrain adaptation futures by reinforcing dominant political-economic structures and processes, and narrowing option spaces; this leads to maladaptive pathways that preclude alternative, locally relevant and sustainable development initiatives and increase vulnerabilities'. The profound transformations needed, it concludes, 'call for examining the values, ethics, attitudes and behaviours that underpin societies. Infusing values that promote sustainable development, overcome individual economic interests and go beyond economic growth, encourage desirable and transformative visions, and care for the less fortunate is part and parcel of climate-resilient and sustainable development pathways' (ibid.: 475). Ecovillages provide spaces where such transformative visions beyond business-as-usual scenarios can be promoted and implemented, and rich lessons learnt.

## A Project in Progress

The Cloughjordan Ecovillage project emerged 20 years ago among a group of pioneers around the Dublin Food Co-op who were concerned at how unaware the public was about climate change impacts and the need to build sustainability. They wanted to show what could be done by developing Ireland's first ecovillage. The project primarily addressed global concerns about climate change and how to address them at a local level. Furthermore, the ecovillage seeks to model economic sustainability and show how this can contribute to social regeneration, especially in rural areas.

The first step was to establish a company in 1999, Sustainable Projects Ireland Ltd. (SPIL),[1] with the overall goal of creating and managing a sustainable village in Ireland, as a model for sustainable living into the twenty-first century and as an education, enterprise, and research resource for all. All ecovillage residents are members, the company is a not-for-profit co-operative, and all decisions are made by members at monthly meetings. Subsidiary goals, as listed in the company's Constitution, are:

1) To minimise pollution to the air, water, and land;
2) To demonstrate a new approach to rural regeneration;
3) To maximise the potential for earning a living, both inside and outside the village, by facilitating systems whereby people can create local and sustainable work;
4) To provide for the cultural, artistic, and non-material needs of the residents of the village and surrounding community.

A 67-acre site behind the main street of Cloughjordan was identified in 2003 and bought in 2004; a plan for a residential area of 132 housing units was drawn up and outline planning permission received from North Tipperary County Council. Finance was raised from members, from an ethical investor and by borrowing. Infrastructure was installed in 2007–08 and building began in 2009. Once Cloughjordan was chosen, a participatory process was initiated with residents of the existing village to imagine how the ecovillage would look and would integrate into the community. Eventually, 55 houses were built by 2013, but 28 site owners have not been able to build and 47 sites remain for sale. A district heating system was installed to heat all the homes in a carbon-neutral way, and a green enterprise centre was built in 2011. Cloughjordan Community Farm was established in 2008 as a Community Supported Agriculture (CSA) farm to develop a resilient food system. In 2014, the ecovillage commissioned Tipperary Energy Agency to measure its ecological footprint (EF) using a methodology based on that of the Global Footprint Network. The average ecovillager's EF of 2 global hectares (gHa) compares favourably to an EF of between 2.9 and 4.3 for other Irish towns and an EF of 5.2 for the average Irish person, as measured by the Global Footprint Network. This means that living as an average ecovillager requires 1.1 Earths, whereas living as an average Irish person requires 3.14 Earths (www.globalfootprintnetwork.org).

The major recession that hit the Irish economy from 2008 onwards, centred on the housing market, had a severe impact as the sale of sites halted. This eroded the company's financial basis, forcing it to cut costs drastically to survive. Despite this, the ecovillage has strengthened its identity as a unique networked hub of organisations, sometimes referred to as 'an ecosystem of innovation', thereby helping to fulfil the objectives it sets itself. Among these are Cloughjordan Community Farm, one of Ireland's most developed CSAs, and Cloughjordan Community Arts, which has built an amphitheatre within the ecovillage. The amphitheatre was opened by President Michael D. Higgins in April 2017. Cloughjordan Community Cooperative, founded in 2016, uses the food produced by Cloughjordan Community Farm and the bread baked by RiotRye bakery in the ecovillage to produce wholesome meals in the co-op's Middle Country Café on Cloughjordan's main street. Cloughjordan Co-Housing is another pioneering co-operative, planning a co-housing development within the ecovillage to offer low-cost and low-energy homes with shared services. Among the micro-enterprises that flourish within this ecosystem are Django's Ecohostel; the WeCreate green enterprise centre which incorporates a 'fab lab'; VINE, which provides internet services in the ecovillage; and RiotRye bakery and baking school. These enterprises co-operate to enhance their businesses and model a form of resilient economic and social regeneration. As an aid to biodiversity, 17,000 native trees were planted as a wild woodland area. As a result of a disease that affected the ash trees in 2016, a major replanting took place in 2018, this time including varieties of oak trees. A labyrinth in the woodland area provides a space for meditative walking. The ecovillage is committed to promoting the Irish language and all street names are in Irish only. Various activities through Irish, including a regular PopUp Gaeltacht, are a feature of the community's life.

Cloughjordan has also gained national and international recognition. It won the National Green Award for Ireland's greenest community three years in a row from 2012 to 2014 and the ecovillage won a gold medal award at the 2013 International Awards for Liveable Communities (LivCom), also known as the Green Oscars; this was hosted by Xiamen in the People's Republic of China and supported by the UN Environment Programme (UNEP). The Milesecure consortium of 14 European research centres identified Cloughjordan Ecovillage as 1 of the 23 most successful models of transition to a low-carbon society out of the 1400 examined in Europe. It was recognised as one of Europe's most innovative social projects in a survey conducted by the Young Foundation in London for the European Commission. Cloughjordan has been ranked by readers of *The Irish Times* in a national survey as one of the ten best places to live in Ireland. It features regularly in radio and TV programmes and numerous articles have been published in the national media on the project.

## Literature Review on Cloughjordan Ecovillage

A significant academic literature has developed on Cloughjordan Ecovillage since building began in 2009. This examines various aspects of the ecovillage project including governance structures and decision-making, the project's food system, sustainable housing, the ecovillagers' ecological footprint, and the collaborative economy of the ecovillage. Papadimitropoulos (2018) examines Cloughjordan Ecovillage as an example of a collaborative economy addressing the devastating consequences for nature and society of a predatory capitalism by introducing new and radical forms of ownership, governance, entrepreneurship, and financialisation to promote sustainability, decentralisation, democratic self-governance, and equitable distribution of value. Following three months of fieldwork, he found that Cloughjordan Ecovillage 'is a notable case of a strong sustainability approach, which, contrary to the dominant neoclassical paradigm, combines individual and collective interest with the aim to tackle climate change on the basis of community living' (58).

In researching how the liveable space of the ecovillage was created, Ranz Mc Donald (2019) examined agency in the project, who and why certain people have influence or agency in the development of space. She found that conflicts in the ecovillage community and an inability to make and enforce decisions within the internal voluntary governance structure (the Viable Systems Model) have caused a partial breakdown of the governance structure and community cohesion. This is manifest in its space and the author concludes that the 'objective liveability' of Cloughjordan Ecovillage is certainly below average, as 'major investments in both time and capital are required to bring this estate's physical space above board'. However, the 'subjective liveability' is high and the environmental impact is far lower than the national average.

Ranz Mc Donald's research updates earlier research on governance systems and practices in the ecovillage. Espinosa and Walker (2013) describe an action research project in which the authors facilitated a process of self-organisation in the ecovillage through the introduction of the Viable System Model (VSM), helping the community identify their primary tasks and develop meta-systematic management tools to deal with complexity. Observing the dynamics of the self-organising process over three years, the authors conclude that roles and tasks were designed more effectively, strengthening viability and sustainability. Examining the consensus decision-making model of the ecovillage, Cunningham (2014) concluded that ‘despite the impressive nature of the built infrastructure at this site, the community continues to struggle with governance, decision-making, consensus and communication issues’. Salter’s (2017) research on socio-psychological sustainability in Cloughjordan Ecovillage found that personal variations in perceptions of Sense of Community (SoC) among ecovillagers were often attributed to ideological dissonance and the incomplete nature of the governance process. She concluded that community should therefore be regarded as a continual process and that component SoC factors must be continually worked upon, through a range of projects and initiatives, to ensure and enhance socio-psychological sustainability in the ecovillage.

Examining Cloughjordan Ecovillage as a case study in sustainable housing, Winston (2012) concludes that it fulfils many of the criteria for sustainability, noting its access to public transport; on-site or local employment; the use of sustainable housing designs and materials; on-site renewable energy; energy-efficient buildings; access to green spaces for food, energy, and leisure; and access to a wide range of social resources. However, she noted that little explicit sustainability assessment had been conducted for the entire enterprise. To assess the ecovillage’s claim to be modelling the transition to a low-carbon society, a process of measuring the average Ecological Footprint (EF) of ecovillagers was undertaken in 2014–2015. Carragher and Peters (2018) describe the process and outline how a bottom-up compound EF method was developed for the purpose. The authors emphasise the purpose of providing meaningful feedback and reflective learning on human carbon intensity to ecovillage residents.

Early in the building of the ecovillage, Campos (2013) examined the challenge of implementing the project’s vision of becoming a self-reliant, socially supportive, and ecologically sustainable community, identifying some challenges of translating this vision into reality. He found that ‘the project hasn’t been able to get around the predatory price system of the formal economy’ and that this constitutes ‘a serious limitation to the development of the enterprise – it potentially excludes members with skills, knowledge and experience that otherwise would be able to give input and boost CEV’s aspirations’ (Campos 2013: 40). He concludes: ‘The paradox then becomes apparent—CEV’s enterprise conveys a vision of an alternative way of living although it runs on, somewhat, mainstream tracks; that is, it is supported by mainstream institutions such as debt money or private ownership' (ibid.: 41). A similar point was identified by Nelson (2018) in her case study of the ecovillage. She concluded that Cloughjordan 'replicates many features found in other ecovillages' such as non-hierarchical work planning and monitoring teams, a modified form of consensual decision-making, and discussion through deep listening and engagement (148). However, of the four ecovillages she studied, she concluded that Cloughjordan 'is the most market-oriented … and the least affordable' (ibid.: 157).

Kirby (2016) gives an overview of how Cloughjordan Ecovillage attempts to model the transition towards a low-carbon society and to disseminate the lessons learnt. He analyses the elements that support modelling the transition, including energy, food (community farm), transport, water, the building of homes, livelihood styles, and community life. The author outlines the ecovillage governance model and how it evolved since inception and describes the ecovillage's educational activities through which the lessons learnt are disseminated.

Milesecure, an EU-funded research project designed to identify the obstacles to, and drivers of, the transition to a low-energy society, undertook an analysis of 1400 projects relating to energy transition at the local level throughout Europe 'so as to identify actual conditions that facilitate (or, conversely, hinder) a transition towards a post-carbon (or low-carbon) society' (Milescure 2014: 7). Of the 90 anticipatory experiences initially identified, 23 were selected for a more detailed examination on the basis of a number of criteria including the success of the project 'in terms of social recognition and excellent results (indicatively defined "post carbon") from an energy and environmental point of view' (15). Among those selected was Cloughjordan Ecovillage. The project's final manifesto highlighted Cloughjordan Ecovillage 'as an example of how a local community can take control over the implementation of low-carbon footprint living areas … a great counterbalance for top-down styles of government' (Walkenburg et al. 2015: 12).

The Atlas of Social Innovation (2018), a global map of social innovation projects produced as part of the EU/FP7 SI-Drive project, a consortium of 23 global partners, identifies Cloughjordan Ecovillage as one of the more than 1000 social innovations featured. Categorised as an 'energy innovation', the Atlas identifies the most important factors accounting for its success as being the leadership of key individuals, EU funding for its district heating system, and what it calls 'the sustainability movement' as it energised people to get involved. While identifying a lack of capacity due to the need to get more young people involved, the Atlas also states that 'there are distinct competencies that could be shared in order to diffuse at least the principles of the innovation more broadly', mentioning consensus decision-making and the viable systems model.

Atlas of Utopias (2019) features Cloughjordan Ecovillage in its energy and food sections. Following a description of the project, it offers the following evaluation: 'The project is conceptually well conceived and implemented in a practical down to earth manner. The presentation is modest and self-critical. It is being implemented with stamina and dedication by the collective members. It is a useful well-documented real-life experiment meriting further study'.

Sempere (2018) examines Cloughjordan Ecovillage as a transition town, looking in particular at its community cohesion and territorial relocalisation. He concludes that practical projects like Cloughjordan 'help to imagine a future of scarcity and prepare better for it. … Its existence is pedagogical: it stands as a sort of "instructions manual" available for problematic situations of scarcity' (120).[2]

Casey et al. (2016) examine sustainable consumption practices in Cloughjordan Ecovillage as a contribution to the shift in sustainable consumption research from a focus on individual behaviours to wider social networks and material infrastructures. They conclude that ecovillage members employ tactics that encourage reflexivity in everyday practices so as to confront routine consumption and create alternative infrastructures that support sustainability and foster critical engagement. Moore et al. (2014) examine a very different form of reflexive practice in Cloughjordan Ecovillage, tracing the development of Cloughjordan Community Farm. They describe how this member-owned and -operated CSA critically self-assessed and restructured in the face of challenges, thereby developing an adaptive awareness which they describe as a 'reflexive resilience'.

## What Makes Cloughjordan an Ecovillage?

Three key features of the Cloughjordan project support its status as an ecovillage: its ecological building standards, its carbon-neutral District Heating System (DHS), and its food system centred on Cloughjordan Community Farm (CCF).

1) Ecological building standards: While the ecovillage does not exercise any control over building designs (which require planning permission from Tipperary County Council), it has set out a range

of standards and principles to which all buildings must adhere. These are contained in the Ecological Charter which sets targets for total heat input and exact specifications for insulation as well as covering issues of air-tight construction, ventilation, and maximising natural light and heat through building south-facing. It recommends the use of non-toxic materials, regionally sourced and with low embodied energy. Guided by the Charter, CEV has pioneered the use of different building types such as passive timber-frame, Durisol blocks of chipped waste wood bonded with ecocement, cellulose, hemp-lime, and cob. In 2017, 10 of the 55 houses in the ecovillage installed photovoltaic panels to generate energy.

2) District Heating System: All homes in the ecovillage depend on the district system for the supply of hot water and heating. The heating plant contains two 500-kilowatt wood-chip boilers backed up by 500 $m^2$ of solar (thermal) panels. This system is the first of its kind in a private housing development in Ireland and is estimated to save some 113.5 tonnes of carbon emissions annually over what would be emitted by an equivalent size development using conventional heating methods. The plant supplies hot water daily to all homes via a well-insulated network of piping and the water is stored in each house in an insulated storage tank supplying hot water and heating.
3) Food system: The CEV food system is centred on Cloughjordan Community Farm, an area of around 10 acres of which about half are under cultivation at any one time, the remainder being under green manures or in pasture. It is run on organic principles and produces some 85 varieties of vegetables, salads, herbs, and fruits to feed a membership of around 90 people, to supply the Middle Country Café co-operative, and to feed visiting groups at events and conferences. It includes a dedicated composting area, a composting toilet, and a compost-heated shower. In addition, a timber building has been erected by farm members and volunteers for seed saving and the social use of the farm workers. The farm is run by two full-time farmers but, as its ecologically sensitive methods are quite labour-intensive and there is minimal use of mechanical equipment, its smooth functioning also relies upon the manual labour of eight volunteers from the European Solidarity Corps (ESC), who each come for a year. CCF now exceeds the United Nations terms for the definition of agro-restorative farming. It is committed to seed saving, and currently about 60% of crops are grown from its own seed.

In these ways, CEV is pioneering a range of practices that build resilience and sustainability. The benefits were illustrated during the severe snowstorm that hit Ireland in early March 2018 known as the 'Beast from the East', when communities throughout the country were threatened by food shortages, interruption of heating systems, and isolation. CEV residents were able to depend on reliable heating, food deliveries from CCF and RiotRye bakery, and a supportive community.

## Facing the Challenges of Building Community

Building CEV has required pioneering new technologies and ecological building techniques, but it has been learnt that building a resilient community is the greatest challenge. This approximates to the growth phases of any individual's life as early idealism meets the severe challenges of implementation and of moving through difficult times. In the case of the ecovillage, the acute recession of 2008, centred on the building sector, had a major impact on the project, resulting in some being unable to move forward with their building plans and a halt to site sales. As a result, SPI had to let go all its staff and the project became entirely dependent on voluntary labour. All members pledge at least 100 hours voluntary labour a year so that the project can draw on an extensive range of skills and experience. These skills and experiences include running SMEs; primary-, secondary-, and third-level education as well as the non-formal sector; organic farming and seed saving; land management and forestry; low-energy building and renewable energy systems; urban drainage systems; waste disposal systems; cooking and nutrition; arts and crafts; health-care and well-being; psychotherapy and spirituality; communications and photography; and social media. Members step up to various roles, on various boards, on the office team, in looking after the land, and in varied educational activities.

From the beginning, CEV adopted a consensus form of decision-making. The monthly members' meeting is the central forum for accountability and for making decisions. This has helped to make members feel they are driving the project. However, its growing complexity required a transformation of governance structures from traditional top-down structures (by 2007, there were 21 working groups) to a flatter and more bottom-up model. Consultants Jon Walker and Angela Espinosa helped design these structures in 2007 using a Viable Systems Model (VSM), originally developed by Stafford Beers in England. Based on Beers' observations of natural eco-systems' self-organising, he designed a layered governance system which gives priority to an organisation's primary activities and identifies the tasks needed to support these activities and help them flourish.

In the case of CEV, three primary activities, or PAs, were identified—education, land use and the development of the built environment. These PAs are organised with multiple internal groups to support the range of tasks and responsibilities held by the PA. Supporting these are a Process group which organises meetings and decision-making processes, the SPI Board which has legal and financial oversight of the project, and numerous other groups looking after such tasks as maintenance and fundraising. The co-ordination function plays a key role as it ensures transparency and accountability to members through a bi-monthly Co-ordination Report to which all groups contribute and which is sent to all members before the members' meeting. Co-ordinators of each group attend the meeting to answer questions and hear the concerns of members. This system has proven very successful in fostering bottom-up creativity throughout the project, but it is found by some to be complex and slow-moving in making decisions. The experience of managing this system has helped to build skills of conflict resolution, facilitation, mediation, and community building, and courses are occasionally provided to train members in these skills. In 2020, the community began to introduce decision-making practices drawn from Sociocracy.

Despite this, there remain some deep divisions in the community. In her research, Ranz Mc Donald (2019) identifies a number of polarising issues. A central one derives from the consequences of establishing a Service Company (SC) in 2007, as required by Tipperary County Council as a condition for granting planning permission. There is disagreement over whether the Multi-Unit Development Act of 2011 applies to the ecovillage and whether ownership of land should be transferred from SPI to the SC so that the latter has legal responsibility for the management of the services on that land, and disagreement over the extent of the land to be transferred. An additional complicating factor is disagreement over who should be members of the project, whether it should be limited to those who have built homes or who rent in the ecovillage, or should also incorporate a wider group of activists and supporters. One resident is quoted as saying that 'there are some very entrenched positions and it is very much tied up with attitudes to land ownership and property ownership' (Ranz Mc Donald 2019: 34). Ranz Mc Donald writes: 'Because of conflicts, social cohesion is nuanced and as a result the community which exists in CEV is more aptly described as a "community of communities"' (ibid.: 45).

## Educating for Sustainability: Sharing the Lessons

Village Education, Research and Training (VERT) is the educational arm of CEV. Its remit is to establish the ecovillage as a leading national and international centre for education on sustainability, resilience, community living, and rural regeneration and for rethinking and remodelling the society of the future. It offers a place-based learning experience of building a sustainable community which is unique to Ireland. This it does mainly through guided tours every Saturday and Sunday and visits by schools, community groups, and specialist professional groups from Ireland and abroad. Longer visits, particularly by groups of third-level students as part of their academic study, involve a wider range of inputs and activities, tailor-made for the needs and interests of each group. VERT is also involved with other Cloughjordan-based learning providers in organising a number of festivals and workshops each year around particular issues such as co-housing, community energy, organic farming, and permaculture. VERT currently has partnership agreements with Dublin City University and with Mount Temple Comprehensive School, Dublin. The appointment of a part-time education officer in 2018 has helped to coordinate VERT's educational offering and to respond more professionally to increasing demand.

Other educational providers are also active in CEV. Cultivate, a national environmental non-governmental organisation (NGO) providing education in sustainability, is headquartered in CEV and collaborates actively with VERT. RiotRye runs courses in natural baking which incorporate information about the ecovillage. Individual residents are also active in providing education on issues related to sustainability.

## Conclusion: Looking to the Future

As SPI celebrated the 20th anniversary of its foundation in 1999 and the 10th anniversary of the first houses being built in 2009, it was planning a new development phase. This involves building out the ecovillage, bringing in new members and broadening the diversity of the community through including social housing and affordable rental options, and developing it as a leading campus for education in all aspects of sustainability.

As Ireland's first ecovillage, it has established itself as a living model of both environmental and economic sustainability, of rural regeneration, and of resilient community. It offers a unique example of place-based learning about how to address the challenges of climate change and biodiversity loss in ways that could be replicated in both urban and rural areas. It pioneered high ecological standards in its building regulations which national regulations were later to catch up with. Its carbon-neutral district heating system offers another model that could be replicated in estates and apartment blocks throughout the country. Its member-based community farm supports a food system which demonstrates an alternative model of agriculture so badly needed in Ireland. This shows how communities could meet more of their own food needs, regenerate the soil, create livelihoods, and eat in a far more healthy way. In these ways, the ecovillage has been a pioneering project in a country that for long neglected to take seriously the challenges of climate change and biodiversity loss. As national policy begins finally to grapple with the immense challenges of social, economic, and cultural transformation required to achieve low-carbon development, the challenge for projects like Cloughjordan Ecovillage is to remain ahead of the curve, learning and spreading the lessons of how vibrant, resilient, and welcoming communities will be the seedbeds of a low-carbon Ireland.

## Notes

1. Known since 2015 as Sustainable Projects Ireland CLG (SPI).
2. Translation from the Spanish by Peadar Kirby.

## Bibliography

*Atlas of Social Innovation*. 2018. Available at: www.socialinnovationatlas.net

*Atlas of Utopias*. 2019. Available at: www.transformativecities.org

Campos, P. 2013. Striving Intentionalities: Vision and Practice in Cloughjordan Eco-Village. MSc dissertation, University of Lund.

Carragher, V., and M. Peters. 2018. Engaging an Ecovillage and Measuring Its Ecological Footprint. *Local Environment: The International Journal of Justice and Sustainability* 8: 861–878. https://doi.org/10.1080/13549839.2018.1481021.

Casey, K., M. Lichrou, and L. O'Malley. 2016. Unveiling Everyday Reflexivity Tactics in a Sustainable Community. *Journal of Micromarketing* 37 (3): 227–239. https://doi.org/10.1177/0276146716674051.

Cunningham, P. 2014. Exploring the Efficacy of Consensus-Based Decision-Making. *International Journal of Housing Markets and Analysis* 7 (2): 233–253. https://doi.org/10.1108/IJHMA-06-2013-0040.

De Schutter, O. 2014. The EU's Fifth Project: Transitional Governance in the Service of Sustainable Societies. Available at: http://www.srfood.org/images/stories/pdf/otherdocuments/Framing4.pdf. Accessed 5 June 2019.

Espinosa, A., and J. Walker. 2013. Complexity Management in Practice: A Viable System Model Intervention in an Irish eco-community. *European Journal of Operational Research* 225: 118–129.

Hickel, J. 2019. Degrowth: A Theory of Radical Abundance. *Real-World Economics Review* 87: 54–68.

Jackson, R., and H. Jackson. 2015. The History of GEN. In *Ecovillage: 1001 Ways to Heal the Planet*, ed. K. Joubert and L. Dregger, 217–227. Devon: Triarchy Press.

Kirby, P. 2016. Cloughjordan Ecovillage: Modelling the Transition to a Low-Carbon Society. In *Transitioning to a Post-carbon Society: Degrowth, Austerity and Wellbeing*, ed. E. Garcia, M. Martinez-Iglesias, and P. Kirby, 183–205. Basingstoke: Palgrave Macmillan.

Litfin, K.T. 2014. *Ecovillages: Lessons for Sustainable Community.* Cambridge: Polity Press.

Milescure-2050. 2014. *Deliverable 2.2: Report on Comparative Analysis.* Available at: www.milesecure2050.eu

Moore, O., O. McCarthy, N. Byrne, and M. Ward. 2014. Reflexive Resilience and Community Supported Agriculture: The Case that Emerged from a Place. *Journal of Agriculture, Food Systems and Community Development* 4 (3): 137–153. https://doi.org/10.5304/jafscd.2014.043.007.

Nelson, A. 2018. *Small Is Necessary: Shared Living on a Shared Planet.* London: Pluto Press.

Papadimitropoulos, V. 2018. Sustainability and Resilience in the Collaborative Economy: An Introduction to the Cloughjordan Ecovillage. *Journal of Public Policy and Administration* 2 (4): 49–60. https://doi.org/10.11648/j.jppa.20180204.12.

Ranz Mc Donald, C. 2019. The Production and Liveability of Space in Cloughjordan Ecovillage. MSc thesis, University of Amsterdam.

Roy, J., P. Tschakert, H. Waisman, W. Abdul Halim, P. Antwi-Agyei, P. Dasgupta, B. Hayward, M. Kanninen, C. Okereke, P.F. Pinho, K. Riahi, and A.G. Suarez Rodriguez. 2018. Sustainable Development, Poverty Eradication and Reducing Inequalities. In *Global Warming of 1.5°C. An IPCC Special Report on the Impacts of Global Warming of 1.5°C Above Pre-Industrial Levels and Related Global Greenhouse Gas Emission Pathways, in the Context of Strengthening the Global Response to the Threat of Climate Change, Sustainable Development, and Efforts to Eradicate Poverty*, ed. V. Masson-Delmotte, P. Zhai, H.-O. Pörtner,

D. Roberts, J. Skea, P.R. Shukla, A. Pirani, W. Moufouma-Okia, C. Péan, R. Pidcock, S. Connors, J.B.R. Matthews, Y. Chen, X. Zhou, M.I. Gomis, E. Lonnoy, T. Maycock, M. Tignor, and T. Waterfield. Geneva: IPCC.

Salter, E. 2017. An Imagined Utopia: An Analysis of Socio-Psychological Sustainability at Cloughjordan Ecovillage. BA dissertation, University of Nottingham.

Sempere, J. 2018. *Las Cenizas de Prometeo: Transición energética y socialismo.* Barcelona: Pasado y Presente.

Sustainable Projects Ireland. 2019. *Cloughjordan Ecovillage Annual Report 2018.* Cloughjordan: SPI.

Walkenburg, C., W.E. Bijker, and T.E. Swierstra, eds. 2015. *A Manifesto for Human-Based Governance of Secure and Low-Carbon Energy Transitions, Deliverable 5.1 of the Milesecure-2050 Project.* Maastricht/Turin: Maastricht University/Politecnico di Torino.

Winston, N. 2012. Sustainable Housing: A Case Study of the Cloughjordan Eco-Village, Ireland. In *Enterprising Communities: Grassroots Sustainability Innovations (Advances in Ecopolitics, Vol. 9)*, ed. A. Davies, 85–103. Bingley: Emerald Group Publishing Limited. https://doi.org/10.1108/S2041-806X(2012)0000009008.

# Index[1]



[1] Note: Page numbers followed by 'n' refer to notes.



D. Robbins et al. (eds.), *Ireland and the Climate Crisis*, Palgrave Studies in Media and Environmental Communication,
https://doi.org/10.1007/978-3-030-47587-1

**D**

The manufacturer's authorised representative in the EU is Springer Nature Customer Service Centre GmbH, Europaplatz 3, 69115 Heidelberg, Germany. If you have any concerns regarding our products, please contact ProductSafety@springernature.com

Printed and bound by CPI Group (UK) Ltd, Croydon, CR0 4YY

30/09/2026

02225494-0009